HAZARDOUS CLASSIFIED LOCATIONS
Electrical Design & Installation

Richard E. Loyd

Delmar Publishers
I(T)P An International Thomson Publishing Company

Albany • Bonn • Boston • Cincinnati • Detroit • London • Madrid • Melbourne
Mexico City • New York • Pacific Grove • Paris • San Francisco • Singapore • Tokyo
Toronto • Washington

NOTICE TO THE READER

Publisher does not warrant or guarantee any of the products described herein or perform any independent analysis in connection with any of the product information contained herein. Publisher does not assume, and expressly disclaims, any obligation to obtain and include information other than that provided to it by the manufacturer.

The reader is expressly warned to consider and adopt all safety precautions that might be indicated by the activities described herein and to avoid all potential hazards. By following the instructions contained herein, the reader willingly assumes all risks in connection with such instructions.

The publisher makes no representation or warranties of any kind, including but not limited to, the warranties of fitness for particular purpose or merchantability, nor are any such representations implied with respect to the material set forth herein, and the publisher takes no responsibility with respect to such material. The publisher shall not be liable for any special, consequential, or exemplary damages resulting, in whole or part, from the readers' use of, or reliance upon, this material.

Cover Design by Mick Brady

Delmar Staff
Publisher: Robert Lynch
Acquisitions Editor: Paul Shepardson
Developmental Editor: Jeanne Mesick
Project Editor: Patricia Konczeski
Production Coordinator: Karen Smith
Art and Design Coordinator: Cheri Plasse

COPYRIGHT © 1997
By Delmar Publishers
an International Thomson Publishing Company

The ITP logo is a trademark under license.

Printed in the United States of America

For more information, contact:
Delmar Publishers
3 Columbia Circle, Box 15015
Albany, New York 12212-5015

International Thomson Publishing Europe
Berkshire House 168-173
High Holborn
London, WC1V 7AA
England

Thomas Nelson Australia
102 Dodds Street
South Melbourne, 3205
Victoria, Australia

Nelson Canada
1120 Birchmont Road
Scarborough, Ontario
Canada, M1K 5G4

International Thomson Editores
Campos Eliseos 385, Piso 7
Col Palanco
11560 Mexico D F Mexico

International Thomson Publishing GmbH
Konigswinterer Strasse 418
53227 Bonn
Germany

International Thomson Publishing Asia
221 Henderson Road
#05-10 Henderson Building
Singapore 0315

International Thomson Publishing—Japan
Hirakawacho Kyowa Building, 3F
2-2-1 Hirakawacho
Chiyoda-ku, Tokyo 102
Japan

All rights reserved. No part of this work covered by the copyright hereon may be reproduced or used in any form or by any means—graphic, electronic, or mechanical, including photocopying, recording, taping, or information storage and retrieval systems—without the written permission of the publisher.

1 2 3 4 5 6 7 8 9 10 XXX 02 01 00 99 98 97 96

Library of Congress Cataloging-in-Publication Data
Loyd, Richard E.
    Hazardous classified locations: electrical design & installation /
  Richard E. Loyd ---- 1996 ed.
      p. cm.
    Includes index.
    ISBN 0-8273-6982-4
    1. Electrical engineering—Safety measures. 2. Explosionproof
electric apparatus and appliances. 3. Industrial buildings—Fires
and fire prevention. I. Title.
TK152.L69 1996
621.319'24'0289—dc20   96-12587
                                                          CIP

DEDICATION

I wish to dedicate this book to my wife, Nancy, who has given me so much encouragement and assistance throughout my career. She's my associate, my girl Friday, my number one, and my biggest fan. Without her, this book as well as the other books would never have been possible. Thanks, Nancy, I love you.

I would also like to dedicate this book to some of the outstanding electricians that I have learned so much from during my career as an apprentice and student, and later as a young journeyman electrician. I have continued to learn from others throughout my career as a contractor, inspector, and code administrator. Today I continue to learn from my peers and students; thank you all. Special thanks go to: Earl Featherston, Rub Cragan, Blaine Hogue, Red Allen, Dale Nieman, and Dean Evans, all in Pocatello, Idaho, IBEW Local Union 449; D. Harold Ware, of Oklahoma City; Bill Harris, Cecil Harrell, Clifford Vann, Bill Taylor, George Ingram, and Larry Sharren, all of Little Rock, Arkansas; Ray Olson, Artie Barker, Gary Gould, and Tom Swinehart (Tom taught me the importance of the arrangement and use of words) of Boise, Idaho. I was reluctant to list these names, because there were so many more that helped me get to where I am today. I never met a person who couldn't teach me something, and I have gained some of my most prized knowledge from some of the most unlikely sources. I can honestly say thanks to everyone I ever met.

CONTENTS

Dedication	iii
Foreword	xi
Preface	xiii
Acknowledgments	xv
About the Author	xvii

CHAPTER 1
Codes and Standards in Wiring and Building Design — 1

Purpose	1
National Fire Protection Association Standards	1
National Electrical Code® (*NEC®*)	2
Using the *NEC®*	2
Other NFPA Standards	3
American National Standards Institute (ANSI)	3
NEMA and IEC Standards	4
Federal Government Standards	4
Local Codes and Requirements	4
Insurance Requirements	5
Nationally Recognized Testing Laboratories (NRTL)	5
Occupational Safety and Health Act of 1970 (OSHA)	7
Review Questions	9

CHAPTER 2
Basic Design Factors — 13

Safety	13
Capacity	14
Flexibility	17
Electrical Systems and Building Fires	17
Fire Resistance and Spread of Fire and Smoke	17
Review Questions	18

CHAPTER 3
Hazardous Locations — 21

NEC® Classifications	23
Classes	23
Divisions	23
Groups	24
Evaluation of Hazardous Areas	24
Classification of Hazardous Areas	24
Electrical Sources of Ignition	25
Look for the Marking or Label	25
Hazardous Locations and the *National Electrical Code®*	28
Explosion-Proof Apparatus	28
Dust-Ignition Proof	28
Hazardous Location	28
Class I	29
Class II	29
Class III	30
Division 1 in the Normal Situation	31
Division 2 in the Abnormal Situation	31
Groups	31
Seals	31
Article 500	31
Classification of Area	32
Review Questions	34

CHAPTER 4
Class I Hazardous (Classified) Areas   37

Class I Locations	37
Class I Location Examples	37
Class I Group Classifications	37
Class I Division 1	38
Examples of Class I Division 1 Locations	38
Class I Division 2	40
Examples of Class I Division 2 Locations	42
Equipment for Class I Division 1 Locations	46
Specific Equipment for Division 1 Locations	47
Equipment for Class I Division 2 Locations	48
Specific Equipment for Division 2 Locations	49
Wiring Methods	49
Class I Division 1	49
Class I Division 2	50
Sealing and Drainage Requirements	51
Drainage	54
Review Questions	56

CHAPTER 5
Class II   59

Class II Locations	59
Class II Location Examples	59
Class II Group Classifications	59
Class II Division 1	60
Class II Division 1 Location Examples	60
Class II Division 2	60
Class II Division 2 Examples	60
Equipment for Class II Locations	62
Specific Equipment for Class II Locations	62
Class II Division 1 Locations	62
Division 1 Grounding Requirements	64
Class II Division 2 Locations	66
Division 2 Grounding Requirements	67
Wiring Methods for Class II Locations	67
Class II Division 1 Wiring Methods	67
Class II Division 2 Wiring Methods	67
Sealing Requirements for Class II Divisions 1 and 2	67
Review Questions	68

CHAPTER 6
Class III   71

Class III Locations	71
Class III Division 1 Locations	71
Class III Division 2 Locations	72
Equipment for Class III Locations	72
Specific Equipment for Class III Locations	72
Class III, Division 1 and 2 Locations	72
Wiring Methods	73
Class III Division 1 Locations	73
Class III Division 2 Locations	73
Grounding and Bonding Class III Divisions 1 and 2	73
Sealing Requirements	74
Review Questions	74

CHAPTER 7
Hazardous (Classified) Locations Specific   77

Commercial Garages, Repair, and Storage	77
Article 500 Areas within 511 Structures	78
Wiring and Equipment	79
Sealing Requirements	80
Wiring above Class I Locations	80
Equipment Located above Class I Locations	81
Ground-Fault Circuit-Interrupter Protection	82
Grounding	82
Review Questions	82

CHAPTER 8
Aircraft Hangars   85

Class I Division 1 Areas	86
Class I Division 2 Areas	86
Nonclassified Areas	86
Wiring and Equipment in Classified Locations	86
Wiring and Equipment Not within Class I Locations	86
Stanchions, Rostrums, and Docks	87
Sealing	87
Additional Specific Requirements	87
Grounding	88
Review Questions	88

CHAPTER 9
Gasoline Dispensing and Service Stations   91

Class I Locations	91
Class I Division 2	92
Areas Not Classified	95
Seals	95
Circuit Disconnects	96
Review Questions	96

CHAPTER 10
Bulk Storage Plants   99

Definition	99
Class I Locations	99
Class I Divisions 1 & 2 Wiring and Equipment	102
Sealing	103
Nonhazardous Locations	103
Grounding	103
Review Questions	103

CHAPTER 11
Spray Application, Dipping, and Coating Processes   105

Defined Location	105
Class I or Class II Division 1 Locations	105
Class I or Class II Division 2 Locations	106
Nonhazardous Locations	107
Wiring and Equipment in Class I Locations	107
Illumination	108
Portable Equipment	110
Special Equipment	111
Wiring and Equipment above Class I and Class II Locations	111
Grounding	111
Review Questions	111

CHAPTER 12
Health Care Facilities, Residential and Custodial Facilities, and Limited Care Facilities   113

Review Questions	114

CHAPTER 13
Article 505 Class I Zone 0, 1, and 2 Locations   117

Zones	117
Groups	118
Equipment	118
Wiring Methods	119
Equipment	119
Grounding and Bonding	119
Summary	119
Review Questions	120

CHAPTER 14
Intrinsically Safe Systems — 123

- What Is Intrinsic Safety? — 123
- Definition — 123
- Wiring Methods — 124
- Sealing — 124
- Identification — 124
- Electrical and Mechanical Separation — 124
- Grounding — 124
- Safety Assured — 125
- Review Questions — 125

CHAPTER 15
Wiring Methods for Hazardous Locations — 127

- Protection against Physical Damage — 127
- Underground Requirements — 130
- Exposure to Different Temperatures — 130
- Expansion Properties of Raceways — 130
- Secure and Support Requirements — 135
- Mechanical and Electrical Continuity — 136
- Induction Heating — 137
- Resistance to Deterioration from Corrosion and Reagents — 138
- Penetration of Fire-Resistive Assemblies — 142
- Combustibility of Raceway Materials and Conductor Insulation — 144
- Wiring in Ducts, Plenums, and Other Air-Handling Spaces — 146
- Raceway Allowable Conductor Fill — 147
- Raceway System — 147
- Advantages of Raceway Systems — 149
- Types of Metallic Wiring Methods Systems — 149
    - *Rigid Metal Conduit (NEC® Article 346)* — 149
    - *Intermediate Metal Conduit (IMC)* — 151
    - *Electrical Metallic Tubing (Article 348)* — 152
- Special Feature Tubular Steel Conduits — 153
    - *Integral Coupling Steel Raceways* — 153
    - *Coated Rigid Conduit & Intermediate Metal Conduit (IMC)* — 153
    - *Liquidtight Flexible Metal Conduit* — 155
    - *Rigid Nonmetallic Conduit Schedule 40 and Schedule 80* — 156
- Review Questions — 158

GLOSSARY — 161

APPENDIX — 163
Hazardous Substances Tables

- Hazardous Substances Flammable Gases and Vapors — 164
- Hazardous Substances Combustible Dusts — 167
- Recommended Uses for Raceways and Other Wiring Methods — 169
- Galvanized Rigid Conduit Specifications — 170
- Intermediate Metal Conduit (IMC) Specifications — 170
- Electrical Metallic Tubing (EMT) Specifications — 171
- Rigid Nonmetallic Conduit (Schedule 40) Specifications — 171
- Rigid Nonmetallic Conduit (Schedule 80) Specifications — 171
- Approximate Diameters for Standard Conduit Locknuts and Bushings — 172
- Recommended Spacings for Conduit and Tubing — 172
- Basic Electrical AC and DC Formulas — 173
- Electrical Symbols — 175
- Chemical Resistance Chart for PVC Conduit & Tubing — 179
- Chemical Resistance Chart for Urethane Coatings — 180
- Chemical Resistance Chart for PVC Coatings — 181

INDEX — 183

FIGURES

Figure F–1	xi
Figure F–2	xii
Figure F–3	xii
Figure 1–1	6
Figure 1–2	6
Figure 1–3	7
Figure 1–4	8
Figure 2–1	13
Figure 2–2	14
Figure 2–3	14
Figure 2–3A	15
Figure 2–3B	16
Figure 2–4	17
Figure 3–1	21
Figure 3–2	22
Figure 3–3	22
Figure 3–4	22
Figure 3–5	25
Figure 3–5A	26
Figure 3–5B	27
Figure 3–6	27
Figure 3–6A	28
Figure 3–7	28
Figure 3–8	29
Figure 3–9	30
Figure 3–10	31
Figure 3–11	32
Figure 3–12	32
Figure 3–13	32
Figure 4–1	37
Figure 4–2	38
Figure 4–3	39
Figure 4–4	41
Figure 4–5	42
Figure 4–6	43
Figure 4–7	45
Figure 4–8	47
Figure 4–9	48
Figure 4–10	48
Figure 4–11	49
Figure 4–12	50
Figure 4–13	50
Figure 4–14	52
Figure 4–15	53
Figure 4–16	53
Figure 4–17	54
Figure 4–18	55
Figure 4–19	55
Figure 4–20	55
Figure 5–1	59
Figure 5–2	60
Figure 5–3	61
Figure 5–4	62
Figure 5–5	62
Figure 5–6	63
Figure 5–7	64
Figure 5–8	65
Figure 5–9	66
Figure 5–10	67
Figure 6–1	71
Figure 6–2	71
Figure 6–3	73
Figure 6–4	74
Figure 7–1	77
Figure 7–2	79
Figure 7–3	80
Figure 7–4	80
Figure 7–5	81

Figure 8–1	85	Figure 15–14	138
Figure 8–2	86	Figure 15–15	139
Figure 8–3	87	Figure 15–16	140
		Figure 15–17	148
Figure 9–1	92	Figure 15–18	149
Figure 9 2	93	Figure 15–19	149
Figure 9–3	95	Figure 15–20	149
		Figure 15–21	151
Figure 10–1	99	Figure 15–22	152
Figure 10–2	100	Figure 15–23	153
Figure 10–3	101	Figure 15–24	154
Figure 10–4	102	Figure 15–25	154
		Figure 15–26	155
Figure 11–1	106	Figure 15–27	156
Figure 11–2	106	Figure 15–28	157
Figure 11–3	107		
Figure 11–4	108	Figure A–1(A)	164
Figure 11–5	108	Figure A–1(B)	165
Figure 11–6	109	Figure A–1(C)	166
Figure 11–7	109	Figure A–2(A)	167
Figure 11–8	110	Figure A–2(B)	168
Figure 11–9	110	Figure A–3	169
		Figure A–4	170
Figure 13–1	119	Figure A–5	170
Figure 13–2	120	Figure A–6	171
		Figure A–7	171
Figure 15–1	128	Figure A–8	171
Figure 15–2	129	Figure A–9	172
Figure 15–3	129	Figure A–10	172
Figure 15–4	129	Figure A–11(A)	173
Figure 15–5	131	Figure A–11(B)	174
Figure 15–6	131	Figure A–12(A)	175
Figure 15–7	133	Figure A–12(B)	176
Figure 15–8	133	Figure A–12(C)	177
Figure 15–9	134	Figure A–12(D)	178
Figure 15–10	135	Figure A–13	179
Figure 15–11	136	Figure A–14	180
Figure 15–12	136	Figure A–15	181
Figure 15–13	137		

FOREWORD

ELECTRIC PER SE

Electricity! In ancient times it was believed to be an act of the gods. Neither Greek nor Roman civilizations understood this thing: electricity. It was not until about A.D. 1600 that any scientific theory was recorded, when after seventeen years of research William Gilbert wrote a book on the subject titled *De Magnete*. It was almost another 150 years before major gains were made in understanding electricity and that it might some day be controlled by man. At this time Benjamin Franklin, the man we often refer to as the grandfather of electricity or electrical science, and his close friend, great historian Joseph Priestley, began to gather the works of the many scientists who had been working independently worldwide over the many years. (One has to wonder where this wondrous industry would be today if the great minds of yesteryear had had the benefit of the great communication networks of today.) Benjamin Franklin traveled to Europe to gather this information, and with that trip our industry began to surge forward for the first time. Franklin's famous kite experiment took place about 1752. Coupling the results of that with other scientific data gathered, he decided he could sell installations of lightning protection to every building owner in the city of Philadelphia. He was very successful in doing just that! Franklin is credited for the terms *conductor* and *nonconductor* which replaced *electric per se* and *non electric*.

With the accumulation of the knowledge gained from the many scientists and inventors of the past, many other great minds emerged and lent their names to even more discoveries. Their names are familiar as other terms related to electricity. Alessandro Volta, James Watt, Andre Marie Ampere, George Ohm, and Heinrich Hertz are names of just a few of the great minds that enabled our industry to come together.

The next major breakthrough came about only one hundred years ago. Thomas Edison, the holder of hundreds of patents promoting DC voltage, and Nikola

Figure F–1 Electricity is here! (Courtesy of Underwriters Laboratories)

Tesla, inventor of the three-phase motor and the promoter of AC voltage, began an inimical competition in the late 1800s. Edison was methodically making inroads with DC current all along the eastern seaboard, and Tesla had exhausted his finances for his experimental project with AC current out in Colorado. The city of Chicago requested bids to light the great "Colombian Exposition of 1892." To light the Chicago world's fair, which celebrated the three hundred years since Columbus's discovery of America, would require building generators and nearly a quarter of a million lights. Edison teamed up with General Electric and submitted a bid using DC, but George Westinghouse and Nikola Tesla submitted the lowest bid using AC and were awarded the contract. It is said that thousands came out to see this great lighting display. But almost as spectacular as the lighting were the arcing, sparking, and fires started by the display. Accidents were occurring daily! It is said that Edison was embellishing the dangers of AC current by electrocuting dogs while promoting DC current as being much safer. But the public knew better since plenty of fires and accidents were also occurring with DC current.

About this time, three very important events took place almost simultaneously. The Chicago Board of Fire Underwriters hired William Henry Merrill, an electrician from Boston, as an electrical inspector for the exposition. Mr. Merrill saw the need for safety inspections and started "Underwriters Laboratories" in 1894. In 1881 the National Association of Fire Engineers met in Richmond, Virginia, and drafted a proposal that became the basis for the *National Electrical Code®*. In 1895 the first nationally recommended code was published by the National Board of Fire Underwriters (now the American Insurance Association). In 1897, using this 1895 code as a basis, the first *National Electrical Code®* was drafted through the combined efforts of architectural, electrical, insurance, and other allied interests. The first *National Electrical Code®* was presented to the National Conference on Electrical Rules composed of delegates from various national associations who voted to unanimously recommend it to their respective associations for adoption or approval. And a group from the city of Buffalo, New York, hired Westinghouse and Tesla to build AC generators for Niagara Falls; AC current had won out over DC current. The rest is history!

Figure F–2 1893 Great Colombian Exposition at the Chicago World's Fair. The Palace of Electricity astonished crowds—it astonished electricians too, by repeatedly setting fire to itself. (Courtesy of Underwriters Laboratories)

Figure F–3 Examples of early equipment labels. (Courtesy of Underwriters Laboratories)

PREFACE

The primary purpose of this book is to provide students specializing in industrial and commercial electrical power and control systems with a concise easily understandable study guide of the classification and installation guidelines for hazardous (classified) locations, the electrical wiring methods, and basic electrical design considerations for hazardous (classified) locations. This text is written recognizing that the electrical field has many facets, and the user may have diverse interests and varying levels of experience. Therefore, in addition to the student, the interests of the electrical designer, the consulting engineer, the installing electrician, and the electrical inspector were carefully considered in the text. It provides a ready source of the basic information on the design of the most common hazardous locations for industrial, institutional, commercial, and residential occupancies. This text may be used by engineers, contractors, and electricians as a reference book. It will also serve as an aid in selecting the appropriate wiring methods for hazardous location installations. In addition to its primary purpose as study material for students and electricians, it is useful for examination preparation.

The text is brief and concise for easy application to everyday practical problems and is supplemented with pictures, drawings, examples, and tables, many of which are extracted from the 1996 *NEC*®. The information contained in this text facilitates efficient, safe, and economical application of the various types of available wiring methods.

For more detailed information on the user's specific needs, reference should be made to the current edition of the *National Electrical Code*®, other National Fire Protection-related Standards, engineering textbooks, application handbooks, and for the appropriate manufacturers' catalogs.

ACKNOWLEDGMENTS

Allied Tube & Conduit Company
Appleton
American Iron & Steel Institute (AISI)
Picoma Industries Inc.
Bussmann Manufacturing Company
Crouse-Hinds Division of Cooper Industries
Lampson & Sessions Company
OCAL INC.
Perma-Cote Industries
Pyrotenax USA Inc.
Red Dot L.F. Mason Co.
Robroy Industries Inc.
Square D Company
R. Stahl Inc.
Triangle Wire & Cable Company
B-Line Systems Inc.
LTV Steel Tubular Products Company
Wheatland Tube Company
Western Tube
Underwriters Laboratories

The author and Delmar Publishers would like to thank those individuals who reviewed the manuscript and offered invaluable suggestions and feedback. Their assistance is greatly appreciated.

Larry Catron
Scott County Vocational School
Gate City, VA 24251

Greg Fletcher
Kennebec Valley Technical Center
Fairfield, ME 04957

Ronald Gossner
Regional Occupational Center
San Jose, CA 95136

Edward Clancy
Sarasota County Technical Institute
Sarasota, FL 34239

Al Gabrysh
St Phillip's College
San Antonio, TX 78203

ABOUT THE AUTHOR

Richard E. Loyd is a nationally known author and consultant specializing in the *National Electrical Code®* and the model building codes. He is president of his own firm, R & N Associates Inc., located in Perryville, Arkansas. He and his wife Nancy travel throughout the country presenting seminars and speaking at industry-related conventions. He also serves as a code expert at thirty-five to forty meetings per year at the International Association of Electrical Inspectors (IAEI) meetings throughout the United States. Mr. Loyd represents the NEMA (National Electrical Manufacturers Association) Section 5RN Steel Rigid Conduit and Tubing as a *NEC®* consultant. He represents the American Iron and Steel Institute (AISI) on ANSI/NFPA 70 the *National Electrical Code®* as a member of Code Committee Eight, the panel responsible for raceways. He is currently chairman of the National Board of Electrical Examiners (NBEE). Mr. Loyd is actively involved in doing forensic inspections and investigations on a consulting basis and serves as an expert witness in matters related to codes and safety.

He is a member of the Arkansas Chapter of Electrical Inspectors Association. He is an associate editor of Intertec Publications (EC&M Magazine). He is also an active member in IAEI (International Association of Electrical Inspectors Association), NFPA (National Fire Protection Association), IEEE (Institute of Electrical & Electronics Engineers Inc.) where he serves on the Power Systems Grounding Committee ("Green Book"), ICBO (International Conference of Building Officials), SBCCI (Southern Building Code Congress International), and BOCA (Building Officials & Code Administrators International). He is currently licensed as a master contractor/electrician in Arkansas (Lic. #1725) and Idaho (Lic. #2077) and is an NBEE-certified master electrician.

Mr. Loyd served as the Chief Electrical Inspector and Administrator for the states of Idaho and Arkansas. He has served as chairman of NFPA 79 "Electrical Standard for Industrial Machinery," as a member of Underwriters Laboratories Advisory Electrical Council, as chairman for Educational Testing Service (ETS) a multi-state electrical licensing advisory board, and as a master electrician and electrical contractor. He has been accredited to instruct licensing certification courses in Idaho, Oregon, and Wyoming and has taught basic electricity and *National Electrical Code®* classes for Boise State University in Idaho.

Chapter 1

Codes and Standards in Wiring and Building Design

PURPOSE

Safety is paramount in an electrical installation where hazardous elements are or may be present. The likelihood that these elements are present and the purpose of the installation must be the prime consideration in designing an electrical system. Over the past 100 years various codes and standards have been developed to provide guidelines and regulations necessary to ensure safe installations. These guidelines govern electricity from its development (generation) and transmission, including associated equipment, to the utilization equipment supplied, and to the maintenance and repair of the electrical system after it is installed. Primary in designing a hazardous wiring system, whether it be a large industrial facility or a very small hazardous material storage closet, is an understanding of the part played by the codes and standards governing each condition. Codes govern what one may or may not do, and standards govern safety and technical aspects of the products used. Standards are also a means of assuring that a variety of parts needed in an electrical system fit together, even when produced by multiple manufacturers. The electrical code is the most important standard for making the electrical installation. Other standards are necessary, however, to properly classify the area within the facility where hazardous elements are or may be present. There are many jurisdictions where the fire and building codes can take precedence and govern wiring decisions. In general, nationally utilized codes and standards are developed by what is known as the "consensus process." As we all know, there can be differing views on how or what we install, or the acceptable level of safety to be achieved. For that reason the consensus method is used, whereby the pros and cons are examined and the language becomes that upon which the majority agrees.

Each organization has its own basic rules and regulations, but the primary goal is to assure that the code or standard under development or revision has widespread distribution and opportunity for review by users, producers, and regulators. The code or standard may carry a dual designation, e.g., ANSI/UL which means Underwriters Laboratories developed the standard, but public review was through the American National Standards Institute.

NATIONAL FIRE PROTECTION ASSOCIATION STANDARDS

The National Fire Protection Association has acted as the sponsor of the *National Electrical Code® (NEC®)* as well as many other safety standards for nearly one hundred years. Each year many standards are revised and new standards are developed to meet the needs of our changing world.

NATIONAL ELECTRICAL CODE® (NEC®)

The most widely used electrical code in the world is the *National Electrical Code®*, whose official designation is ANSI/NFPA 70. The *NEC®* was first developed about one hundred years ago by interested industry and governmental authorities to provide a standard for the essentially safe use of electricity. It is now revised and updated every three years. Each new edition supersedes all previous editions. It was first developed to provide the regulations necessary for the practical safeguarding of persons and property from the hazards arising from the use of electricity. There are about four thousand proposals to amend the code each three-year period. These proposals for change are primarily to introduce new products and to clarify the existing text. The *NEC®* is developed by twenty different code-making panels (CMPs) composed entirely of volunteers. These volunteers come from contractors, installers, manufacturers, testing laboratories, inspection agencies, engineering users, governmental agencies, and the electrical utilities.

Suggestions for the content come from a wide array of sources, including individuals just like you. Anyone can submit a proposal, or make a comment on a proposal that has been submitted by another individual. For more specifics on the *NEC®* process, you may contact the National Fire Protection Association, Batterymarch Park, Quincy, MA 02269. A free booklet, "The NFPA Standards Making System," is available upon request.

The *National Electrical Code®* is purely advisory as far as NFPA and ANSI are concerned but is offered for use in law and regulations in the interest of life and property protection. The name "The National Electrical Code" might lead one to believe that this document is developed by the federal government. This is not so; the *NEC®* only recognizes use of products in a particular way. It approves nothing and has no legal standing until it has been adopted by the *authority having jurisdiction* (see *NEC®* definition Section 90–4), usually a governmental entity (e.g., federal, state or city). Therefore, it is necessary to first check with the local electrical inspection department to see which edition of the *NEC®* has been adopted, if any. Compliance with the *NEC®*, coupled with proper maintenance, should result in an installation essentially free from hazard, but not necessarily efficient, convenient, or adequate for good service or future expansion of electrical use. To meet the specific requirements of commercial and industrial installations, architects and engineers generally specify beyond the minimum requirements of the *NEC®*, particularly in hazardous installations where factors specific to each installation justify the use of methods and materials which will provide greater protection and an additional level of safety. The *National Electrical Code®* is not an instruction manual for untrained persons, nor is it a design specification. However, it does offer design guidelines.

NEC® Section 90–2 defines the scope of the *NEC®*. It is intended to cover all electrical conductors and equipment within or on public and private buildings or other structures, including mobile homes, recreational vehicles, and floating buildings; and other premises such as yards, carnivals, parking and other lots, and industrial substations. The Code also applies to installations of conductors and equipment that connect to the supply of electricity, other installations of outside conductors on the premises, and installations of optical fiber cables. The *NEC®* is not intended to cover installations in ships, watercraft, railway rolling stock, aircraft, automotive vehicles, underground mines, and surface mobile mining equipment. It also is not intended to cover installations governed by the utilities, such as communication equipment, or transmission, generation, and distribution installations, whether in buildings or on right-of-ways. **Note:** For the complete laundry list of exemptions and coverage, see 1996 *NEC®* Section 90–2.

USING THE *NEC®*

Mandatory rules in the *NEC®* are characterized by the word "shall," such as "shall be permitted" or "shall not be permitted." Explanatory information and references are in the form of "Fine Print Notes," which are identified (FPN). All tables and footnotes are a part of the mandatory language. Material identified by the superscript letter "x" includes text extracted from other NFPA documents. A complete list of all NFPA documents referenced can be found in the appendix to the *NEC®*. New revisions to the *NEC®* are identified by: (a) a vertical line is placed beside the added text in the margin where new text has been inserted and (b) where text has been deleted, the deletion is indicated by a "bullet," an enlarged black dot, placed in the margin.

> **Warning:** Where text has been relocated to another section or article, there is no indication given to alert the user that the requirements are still in the *NEC®*.

To use the *NEC®* one must first have a thorough understanding of Article 90 "The Introduction," Article 100 "Definitions," Article 110 "Requirements for Electrical Installations," and Article 300 "Wiring Methods." The next step is to become familiar with Article 230 on services, Article 240 on overcurrent protection, and Article 250 which covers grounding. The rest of the code book can be referred to on an as-needed basis for a particular type of installation.

It is important to remember that one portion of the Code can override another, sometimes without reference. Therefore, it is imperative to look beyond general wiring methods to the specific type of installation. Some examples are Section 300–22 (on plenums and environmental air spaces), Article 517 (on health care facilities), and Article 518 (on places of assembly). **Note:** Other codes and standards may take precedence over the *NEC®*. The NFPA 101 Life Safety Code, applicable building codes, and fire codes may be more restrictive than this article. All of these require specific types of wiring methods under certain conditions.

The installer should verify the environment for the installation before selecting the wiring method. Choosing and installing the wrong method before checking the Code thoroughly may prove to be very costly when an inspector notifies you that the installation has to come out!

OTHER NFPA STANDARDS

Other NFPA standards relating to hazardous electrical installations and their safety are:

NFPA 30	Flammable Liquids Code
NFPA 30A	Automotive & Marine Service Station Code
NFPA 88A	Parking Structures
NFPA 88B	Repair Garages
NFPA 99	Health Care Facilities
NFPA 101	Life Safety Code
NFPA 496	Purged Enclosures
NFPA 497A	Electrical Installations Classification of Class I Hazardous Areas
NFPA 497B	Electrical Installations Classification of Class II Hazardous Areas
NFPA 497M	Electrical Equipment in Chemical Atmospheres

Note: These are only examples of the many NFPA standards that are applicable to specific conditions and installations. There are many other publications which may be necessary in analysis of the conditions and areas to be classified. Many of these publications may be referenced in the appendix of the various NFPA standards, i.e., Occupational Safety and Health Act of 1970 (OSHA), American Petroleum Institute (API), American Society of Testing and Materials (ASTM) Publications, U.S. Bureau of Mines, and National Academy of Sciences. Designers and installers should review the applicable standards before designing or making hazardous installations.

AMERICAN NATIONAL STANDARDS INSTITUTE (ANSI)

The American National Standards Institute (ANSI) is an umbrella organization for the consensus code-making process. ANSI has some of its own volunteer committees which develop standards which are published and sold. In addition, ANSI is an organization that coordinates the efforts of several other organizations which develop codes and standards for the electrical industry; examples are: the National Fire Protection Association (NFPA), the Institute of Electrical and Electronic Engineers (IEEE), and Underwriters Laboratories (UL). After a standard has gone through the ANSI process and received public review, the standard then becomes an "American National Standard." The technical information found in the purely ANSI standards may be identical or similar to that found in the dual-designated UL, *NEC®*, and IEEE standards which are identified ANSI/UL. The designations ANSI/NFPA 70 (the *National Electrical Code®*) or

ANSI/IEEE/ANSI/NEMA indicate the document has been circulated for comment through the ANSI system.

NEMA AND IEC STANDARDS

The National Electrical Manufacturers Association (NEMA) also develops standards for some products. These standards are processed through the ANSI mechanism. NEMA standards are developed by manufacturers' representatives who belong to various NEMA sections (industry groups). They write from intimate knowledge of the products they produce. These standards are to provide product consistency, information, and safety.

Some major NEMA standards deal with switchgear and overcurrent protection devices. The major NEMA raceway standard is RN–1 which covers plastic-coated tubular steel; there is no UL or ANSI standard counterpart for this product. The most common of the NEMA standards are the NEMA enclosure types. Electricians commonly order panelboards, safety switches, and junction boxes by the NEMA enclosure type, such as a NEMA 3R for raintight or a NEMA 1 for an indoor dry location. Enclosures for hazardous locations include the NEMA 7 and the NEMA 10. When these enclosures are properly installed and maintained they are designed to contain an internal explosion without causing external hazard. NEMA Type 7 enclosures are designed for indoor use in locations classified as Class I Groups A, B, C, and D. NEMA type 8 enclosures are designed to prevent combustion through the use of oil-immersed equipment. Type 8 is suitable for use in areas classified Class 1 Groups A, B, C, and D indoor or outdoor. NEMA Type 9 enclosures are designed to prevent the ignition of combustible dust and are suitable for locations classified Class II Groups E, F, and G. NEMA Type 10 are nonventilated types capable of meeting 30 C.F.R. Part 18 of the Mine Safety and Health Administration (MSHA).

NEMA also coordinates much of the input to documents designed by the International Electro-Technical Commission (IEC). This organization is based in Europe and is engaged in writing electrical standards for use around the world. The United States is only one of many countries providing input to these standards and has only one vote. Little by little, IEC standards are making their way into this country. However, they may have some modifications and generally are issued as a UL version of the IEC standard. IEC is primarily mentioned here so you will be aware should the term show up in a specification or as a label on a piece of equipment. It is suggested that you read the magazine articles as they appear on IEC- and EC92-related subjects to gain knowledge of overseas activity which will impact the electrical industry in this country for years to come.

FEDERAL GOVERNMENT STANDARDS

The federal government had its own standards for most products until the late 1970s. In the interest of economics there was a move toward adopting industry standards and canceling the special government specifications, which not only took much time to prepare but created excessive purchase costs because special manufacturing was unnecessarily required. Because many designer specifications still contain outdated references to federal specifications, the old number will be referenced in this book for correlation. For the most part, ANSI and UL standards have now been adopted by the federal government for the wiring systems discussed herein.

LOCAL CODES AND REQUIREMENTS

Although most municipalities, counties, and states adopt the *National Electrical Code*®, they may not have adopted the latest edition, and in some cases the authority having jurisdiction (AHJ) may be several editions of the *NEC*® behind. A majority of the jurisdictions make amendments to the *NEC*® or add local requirements. These may be based on environmental conditions, fire safety concerns, or other local experience. An example of one quite common local amendment is that all commercial buildings be wired in raceways. The *NEC*® generally does not differentiate between wiring methods in residential, commercial, or industrial installations; however, many local jurisdictions do.

Metal wiring methods, especially in fire zones, is another common amendment. Some major cities have developed their own electrical code, e.g., Los Angeles,

New York, Chicago, and several metropolitan areas in Florida. In addition to the *National Electrical Code®* and all local amendments, the designer and installer must also comply with the local electrical utility rules. Most utilities have specific requirements for installing the service to the structure. There have been many unhappy designers and installers who have learned about special jurisdictional requirements after making the installation, thereby incurring costly, and time-consuming corrections at their own expense.

> As an example, when I was Administrator and Chief Electrical Inspector for the State of Idaho, I encountered a situation where the Health Department and EPA required many home owners desiring to build near lakes, rivers, and other waterways to locate their septic systems a prescribed distance away from the water's edge. This was often several hundred feet. To accomplish this generally required the installation of a small sewage holding tank with a sewage pump to move the sewage away from the water to a safe distance where the septic system could be safely located. We recognized that there was a possibility of the accumulation of methane gas in this small pumping station. Because of that possibility, it had to be classified as a Class I Division 1 location. However, to require all homeowners to build or purchase a tank that would meet the requirements of a Class I Division 1 area was very expensive. Therefore, with the cooperation of the Electrical Division of the Idaho Department of Labor, a tank was designed and built that would limit the hazardous area to a small area where only the conductors passed through and contained no splices or equipment. These items were located to provide the degree of safety required at the lowest possible cost.
>
> The pump motor was submersed, and the low-level float switch was located above the top of the pump motor. An additional redundant emergency kill float switch was also located above the top of the motor, but just below the low-level switch. By doing this we permitted a standard pump (not a listed explosion-proof pump). The area above the water level within the tank was classified as a Class I Division 1 location. This area only contained the unbroken cord from the motor. The cord then entered rigid metal conduit or intermediate metal conduit in compliance with Section 501–4(a) an explosion-proof seal was required within 18 inches and splices were made in an explosion-proof box. A general purpose controller was used to control the system.
>
> It is often possible to work together to provide the level of safety necessary by limiting the hazardous area to a minimum and relocating arc-producing equipment out of those areas, thereby eliminating the need for expensive equipment approved for classified locations (see Figure 1–1).

INSURANCE REQUIREMENTS

Many insurance companies have requirements that will affect the insurability and rates of the structure. Those requirements should be determined prior to completing the design stage. The degree of risk to the insurer will be an important determining factor. It may be possible to make design changes which will greatly improve the insurability and rates. The changes necessary to achieve these desired benefits may involve architectural and ventilation changes; therefore, all interested parties should be involved. In some cases, the wiring methods used and the electrical wiring design only may affect such improvement.

NATIONALLY RECOGNIZED TESTING LABORATORIES (NRTL)

Underwriters Laboratories (UL) has long been the major product testing laboratory in the electrical industry. In addition to testing, UL is a longtime developer of product standards. By producing to these standards and contracting for follow-up service after undergoing a listing procedure, a manufacturer is authorized to apply the UL label or to mark the product. In the last decade, numerous other electrical testing laboratories have come on the scene and are being officially recognized. Underwriters Laboratories is not the only testing laboratory evaluating electrical products any more. There are many testing laboratories operating today. Some jurisdictions evaluate and approve laboratories; others accept them based on reputation. OSHA is now evaluating and approving testing laboratories. It is the duty of the entity responsible for specifying the materials to verify that the product has been evaluated by a testing laboratory acceptable to the authority having jurisdiction where the installation is being made. It is the installer's responsibility to

Figure 1–1 Example of a sewage lift pumping station which is acceptable by the authority having jurisdiction.

assure that the product is installed in accordance with the manufacturer's instruction and listing (*NEC*® Section 110–3[b]).

On most products you will find a "Listing Mark" and/or a "Label." Some AHJs will require that you install only listed products in their jurisdiction. Some of the better-known testing agencies are Metropolitan Electrical Testing (MET), ETL Testing Laboratories (ETL), and Factory Mutual (FM). Examples of testing agency trademarks are shown in Figure 1–2.

These testing laboratories generally don't write standards. The testing they perform is performed to UL-, ANSI-, or ASTM-established standards. All except Factory Mutual (FM) have not yet ventured into writing standards.

NEC® Section 110–3(b) requires that you install products in accordance with the manufacturer's instructions, which are part of the listing, and in accordance with the *NEC*®. The instructions may or may not be included with the product (see Figure 1–3). The testing laboratory evaluates products according to a standard which is designed to assure that all manufacturers of a specific item meet a minimum requirement that has been developed based on expected use in accordance with the *NEC*® (see Figure 1–4). Many electrical system failures, accidents, and fires originate from products that were installed incorrectly or were not the correct product for the application, which is a violation of *NEC*® Section 110–3(b). It is interesting to note that although most general purpose motors are *not* UL listed, most motors designed for use in hazardous (classified) locations are UL listed. The UL "Green Book" (*Electrical Constructions Material Directory*), UL "White Book" (*General Information for Electrical*

Figure 1–2 Examples of the trademarks of several well-known testing laboratories.

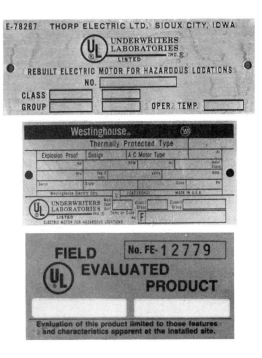

Figure 1–3 Examples of different labels as they appear on equipment. (Courtesy of Underwriters Laboratories)

Construction, Hazardous Location, and Electric Heating and Air-Conditioning Equipment), and the UL "Red Book" (*Hazardous Location Equipment*) will provide valuable information which can help avoid the misapplication of materials. These are the only documents which provide information regarding limitations and specific requirements for a product listing and should be a part of every designer's and installer's library. These and other UL directories and product standards may be obtained from Underwriter's Laboratories Inc., 333 Pfingsten Road, Northbrook, IL 60062.

OCCUPATIONAL SAFETY AND HEALTH ACT OF 1970 (OSHA)

The Occupational Safety and Health Administration (OSHA) began developing and issuing standards intended to protect employees from hazards in the workplace about 1970. In an attempt to eliminate electrical hazards in the workplace, OSHA has continued, and continues today, to develop consensus standards to address specific safety concerns associated with design, installation, maintenance, and repair of electrical systems and components. General safety standards have been developed that cover construction sites and the maintenance and repair of electrical powered machines and equipment.

> Every electrical system, new or existing, must satisfy these requirements. They are fully retroactive, without reference to time. Any violations of these rules constitute OSHA violations and are cause for citations.

All electrical systems must be in complete compliance with all of the rules, or they are in violation of the OSHA law and must be altered or retrofitted to produce full compliance with every one of the OSHA rules. Electrical installations made after March 15, 1972, are covered by this OSHA Act of 1970. All were required to comply with the rules in Section 1910.302 through 1910.308 with eleven specific exceptions. Additional requirements were added effective April 16, 1981. All of the existing electrical installations were required to comply with the rules of the standards 1910.302 through 1910.308 without exception.

Today the new design and safety standards for electrical systems are found in subpart S, Part 1910 of Title 29 of the *Code of Federal Regulations*. The entire

IDENTIFICATION OF UL LISTED PRODUCTS THAT ARE ALSO UL CLASSIFIED IN ACCORDANCE WITH INTERNATIONAL PUBLICATIONS

Underwriters Laboratories Inc. (UL) provides a service for the Classification of products that not only meet the appropriate requirements of UL but also have been determined to meet appropriate requirements of the applicable international publication(s). For those products which comply with both the UL requirements and those of an international publication(s), the traditional UL Listing Mark and a UL Classification Marking, as described below, may appear on the product as a combination Listing and Classification Marking.

The combination of the UL Listing Mark and Classification Marking may appear as authorized by Underwriters Laboratories Inc.

LISTED
(Product Identity)
(Control Number)

ALSO CLASSIFIED BY UNDERWRITERS LABORATORIES INC.® IN ACCORDANCE WITH IEC PUBLICATION _____

LISTED
(Product Identity)
(Control Number)

ALSO CLASSIFIED BY UNDERWRITERS LABORATORIES INC.® IN ACCORDANCE WITH IEC PUBLICATION _____

UND. LAB. INC. ® LISTED
(Product Identity)
(Control Number)

ALSO CLASSIFIED BY UNDERWRITERS LABORATORIES INC.® IN ACCORDANCE WITH IEC PUBLICATION _____

Underwriters Lab. Inc. ® LISTED
(Product Identity)
(Control Number)

ALSO CLASSIFIED BY UNDERWRITERS LABORATORIES INC.® IN ACCORDANCE WITH IEC PUBLICATION _____

(Product Identity)
CLASSIFIED BY
UNDERWRITERS LABORATORIES INC.®
IN ACCORDANCE WITH
IEC PUBLICATION _____
(Control Number)

OR

(Product Identity)
CLASSIFIED BY
UNDERWRITERS LABORATORIES INC.®
IN ACCORDANCE WITH
IEC PUBLICATION _____
(Control Number)

LOOK FOR THE CLASSIFICATION MARKING

(Product Identity)
(Control Number)

ALSO CLASSIFIED BY UNDERWRITERS LABORATORIES INC.® IN ACCORDANCE WITH IEC PUBLICATION _____

(Product Identity)
(Control Number)

ALSO CLASSIFIED BY UNDERWRITERS LABORATORIES INC.® IN ACCORDANCE WITH IEC PUBLICATION _____

UNDERWRITERS LABORATORIES INC. ®
LISTED
(Product Identity)
(Control Number)

ALSO CLASSIFIED BY UNDERWRITERS LABORATORIES INC.® IN ACCORDANCE WITH IEC PUBLICATION _____

LOOK FOR THE COMBINATION LISTING MARK AND CLASSIFICATION MARKING

IDENTIFICATION OF PRODUCTS CLASSIFIED TO INTERNATIONAL PUBLICATIONS ONLY

Underwriters Laboratories Inc. (UL) provides a service for the classification of products that have been determined to meet the appropriate requirements of the applicable international publication(s). For those products which comply with the requirements of an international publication(s), the Classification Marking may appear in various forms as authorized by Underwriters Laboratories Inc. Typical forms which may be authorized are shown below, and include one of the forms illustrated, the word "Classified", and a control number assigned by UL. The product name as indicated in this Directory under each of the product categories is generally included as part of the Classification Marking text, but may be omitted when in UL's opinion, the use of the name is superfluous and the Classification Marking is directly and permanently applied to the product by stamping, molding, ink-stamping, silk screening or similar processes.

Separable Classification Markings (not part of a name plate and in the form of decals, stickers or labels) will always include the four elements: UL's name and/or symbol, the word "Classified", the product category name, and a control number.

The complete Classification Marking will appear on the smallest unit container in which the product is packaged when the product is of such a size that the complete Classification Marking cannot be applied to the product or when the product size, shape, material or surface texture makes it impossible to apply any legible marking to the product. When the complete Classification Marking cannot be applied to the product, no reference to Underwriters Laboratories Inc. on the product is permitted.

Figure 1–4 Identification marks for different types of listings and classifications. This material extracted from Underwriters Laboratories Inc. *General Information for Electrical Construction, Hazardous Location and Electrical Heating and Air-Condition Equipment Directory.* (Courtesy of Underwriters Laboratories)

---

document is divided into major sections and numerically designed as follows:

- 1910.146 Permit Required for Confined Spaces for General Industry. Effective April 1, 1993
- 1910.147 Control of Hazardous Energy (Lockout/Tagout)
- 1910.269 Electric Power Generation, Transmission, and Distribution; Electrical Protective Equipment
- 1910.302 Electrical Utilization Systems
- 1910.303 General Requirements
- 1910.304 Wiring Design and Protection
- 1910.305 Wiring Methods, Components and Equipment for General Use
- 1910.306 Specific Purpose Equipment and Installations
- 1910.307 Hazardous (Classified) Locations
- 1910.308 Special Systems
- 1910.399 Definitions Applicable to this Subpart

The newly adopted 29 CFR 1910.146 sets access to restricted or limited occupancy space areas not designed for continuous occupancy and are hazardous or may be hazardous, i.e., grain silos. These new rules require the employer to set up a sequence of operations, such as:

- Identify hazard(s).
- Identify the presence and availability of oxygen.
- Set up a system to identify each employee entering the work assignments.
- Prevent unauthorized entry.
- Provide necessary protective equipment.
- Provide needed protective clothing.
- Provide at least one worker outside the area to summon help if needed.
- Provide briefing and debriefing to inform all employees and contractors of these procedures.

> The employer is responsible for naming supervisors or designated persons responsible for carrying out each of the tasks.

The OSHA requirements state that all electric products, conductors, and equipment must be approved (listed and/or labeled by an approved testing laboratory NRTL), or they must be replaced with products that are acceptable by OSHA definition, and they must be installed in accordance with the manufacturer's instructions included in the listing or labeling for every installation requiring the use of approved equipment. The rule clearly states that any and every product listed in the UL *Electrical Construction Material Directory* or in any other listing directory must be used exactly as described in the application data given in the listing in the book. OSHA is presently evaluating and approving testing laboratories. It is necessary to consult the Federal Register to determine which testing laboratories (NRTL) are acceptable to the OSHA guidelines.

> The Occupational Safety and Health Administration has announced the availability of an OSHA FAX service. Callers can dial 1 (900) 555-3400 to obtain information on various OSHA regulations. The system offers several hundred documents at a cost of $1.50 per minute, billed on your telephone bill. A Touch-Tone telephone and FAX machine are required to utilize this service. Callers may request a maximum of two documents per call.
>
> The database will contain agency news releases, fact sheets, publication listings, OSHA office listings, inspection statistics, and other information. Most of this information can also be obtained by mail through OSHA Publications, Room N3101, 200 Constitution Ave. NW, Washington, D.C. 21210

REVIEW QUESTIONS

(Answers to some of these questions may be found in the introduction and 1996 *NEC®*.)

1. How often is the *National Electrical Code®* revised and updated?
   Answer: Reference:

2. Who is the authority having jurisdiction?
   Answer: Reference:

3. Name an important item to determine prior to designing or installing an electrical system.
   Answer: Reference:

4. Name four types of installations *not* covered by the *NEC*®.
   Answer: Reference:

5. How is explanatory information characterized in the *NEC*®?
   Answer: Reference:

6. What section of the *NEC*® requires equipment to be installed in accordance with the manufacturer's instructions?
   Answer: Reference:

7. When does the *NEC*® become a legal document?
   Answer: Reference:

8. Are all installations by utilities exempt from the *NEC*®? Explain.
   Answer: Reference:

9. What does a vertical line on the margin of the *NEC*® indicate?
   Answer: Reference:

10. Is the latest amended edition of the *NEC*® used in all jurisdictions? When was the latest edition of the *NEC*® published?
    Answer: (a)   Reference:

               (b)

11. Who invented the three-phase motor?
    Answer: Reference:

12. Who set up the first meeting, that resulted in the development of the *National Electrical Code*®? When and where was it held?
    Answer: Reference:

13. Which organization is engaged in writing electrical standards for use around the world?
    Answer: Reference:

14. Who is responsible for enforcing the OSHA Act of 1970 on construction sites?
    Answer: Reference:

    Noncompliance will result in:
    Answer: (a)

            (b)

15. Where are the OSHA design and safety standards found?
    Answer: Reference:

16. Does the city of Chicago adopt the *National Electrical Code*®?
    Answer: Reference:

17. Which chapters of the *NEC*® apply generally?
    Answer: Reference:

18. Which chapter of the *NEC*® stands alone and is independent of the other chapters?
    Answer: Reference:

19. What does a bullet in the margin of the *NEC*® mean?
    Answer: Reference:

20. Where are commonly-used words or terms defined?
    Answer: Reference:

Chapter 2

Basic Design Factors

Several factors must be considered together in order to achieve a well-designed electrical system for a building or structure (see Figure 2–1). Consideration must be given to the safety requirements. This is particularly important for facilities where hazardous substances are handled or stored. The safety considerations include the wiring methods and architectural considerations, such as means of egress and area separation. The capacity of the electrical service must be considered, as well as voltage regulation. The service should be large enough to handle future expansion, and the voltage should be selected accordingly. Consideration should also be given to accessibility and flexibility. These factors are applicable to the complete distribution or utilization system or part thereof. The ultimate goal is to provide an electrical system that will be efficient to operate, meet initial as well as future requirements, and lend itself to easy maintenance and economical alterations.

Before drawing the plans, a few things must be known so that a system can be designed that will provide "adequacy" (Section 90–1[b]). The *NEC®* states that the minimum requirements of the *NEC®* will provide an installation that is safe and free from electrical hazards, but the installation may not provide all the convenience we would like nor provide provisions for future expansion.

SAFETY

The initial classification prior to the designer developing the plans and specifications provides a foundation upon which a safe electrical installation can be designed and installed. The initial design is based upon the first four chapters of the *NEC®*. The hazardous (classified) areas of the electrical installation require far greater considerations (see Figure 2–2). It is necessary to include the requirements of chapters one through four as applicable. The wiring methods and equipment must be selected based on the requirements in the appropriate article(s) in Chapter 5. Other NFPA standards will need to be used for area classification and to provide guidance in making the final decision on the system design and the selection of equipment and methods. Close cooperation with the authority having jurisdiction, along with others such as the fire marshal and insurance company officials, is needed. The input of a knowledgeable designing engineer is essential

Figure 2–1 Basic design factors must be known before beginning.

> **WHAT ARE HAZARDOUS (CLASSIFIED) LOCATIONS?**
>
> Locations are classified depending on the properties of the flammable vapors, liquids, or gases, or combustible dusts of fibers which may be present and the likelihood that a flammable or combustible concentration or quantity is present.
>
> Each room, section, or area must be individually classified by indicating:
> — Class
> — Division
> — Group

Figure 2–2 Careful consideration must be given and all classifications made prior to completing the design phase.

to ensure that the electrical system incorporates safeguards for the specific conditions. Consideration must be given to the location, occupancy, use, and the possibilities of future expansion. The electrical design should include initial as well as future power capabilities.

NEC® Article 305 and OSHA offer guidelines and requirements for safe, temporary electrical systems and safe working conditions during the installation (construction) stage.

CAPACITY

It is not uncommon for an electrical system to provide continual service for fifty years or more. Therefore, the designed capacity of a new installation must receive comprehensive consideration and will often require close coordination with the serving utility, the structure owner, and the designer. Although oversizing a new system will provide good future expansion capabilities at a relatively low cost, it may not be usable when needed if not closely coordinated with all interested parties (see Figure 2–3).

Special attention should be given to the anticipated needs as well as the future contingency of an electrical service. Consideration should be given to air-conditioning and heating requirements, elevators, data processing and computers, indoor and outdoor lighting, and processing equipment. Future automation and state-of-art upgrading in industrial, commercial, institutional, and residential installations must be considered. It is recommended that service equipment, distribution equipment (switchboards and panelboards), feeders, raceways and conductors, junction boxes and cabinets be specified at least one size larger than the calculated electrical load would require (see Figures 2–3A and 2–3B). Branch-circuit panelboards should contain no less than 10%, preferably 25%, spare circuit breaker or fuse spaces for expansion capabilities. Working space (physical room) required by *NEC®*

Figure 2–3 It makes good sense to consider expansion when specifying the distribution system. (Electrical Switchboard Courtesy of Square D Company)

Examples of Maximum Length Equipment Grounding Conductor (Steel EMT, IMC, GRC, Copper or Aluminum) Computed As A Safe Return Fault Path To Overcurrent Device
Based on 1994 Georgia Tech Software (Version 1.3)
With an Arc Voltage of 40 and 4 IP at 25° C Ambient

Circuit Voltage To Ground	Overcurrent Device Rating Amperes (75° C)	400% (4IP) Overcurrent Device Rating Amperes	Circuit Conductor Size AWG-kcmil Copper or Aluminum	(1) Conduit Trade Size	(2) Equipment Grounding Conductor Size Copper or Aluminum	Length of EMT Run Computed Maximum (In Feet)	Length of IMC Run Computed Maximum (In Feet)	Length of GRC Run Computed Maximum (In Feet)	(2) Copper Grounding Conductor Max Run (In Feet)	(2) Aluminum Grounding Conductor Max Run (In Feet)	(3) REQUIRED CONDUIT SIZE BASED ON NEC 310-16/250-95 & TABLES 1, 4 AND 5 OF CHAPTER 9		
											ENT	SCH 40 PVC	SCH 80 PVC
120	20	80	12	1/2	---	395	398	384	---	---			
120	20	80	12	---	12	---	---	---	300	---	3/4	1/2	3/4
120	20	80	10 AL	---	10 AL	---	---	---	---	293	3/4	3/4	3/4
120	30	120	10	1/2	---	404	(4) 383	---	---	---			
120	30	120	10	---	10	---	399	386	319	---	3/4	3/4	3/4
120	30	120	10	3/4	---	---	(4) 414	---	---	310	1	1	1
120	30	120	8 AL	---	8 AL	447	431	418	---	---			
120	40	160	8	---	---	---	---	---	294	---			
120	40	160	8	1	10	404	400	382	---	232	1	1	1 1/4
120	40	160	8	---	10	---	---	---	228	---			
120	40	160	8 AL	3/4	8 AL	402	397	373	229	221	1	1	1 1/4
120	60	240	6	1	---	---	---	---	---	---			
120	60	240	6	---	10	---	---	---	---	222	1	1	1 1/4
120	60	240	4 AL	---	8 AL	390	388	363	201	---	1 1/4	1 1/4	1 1/2
120	100	400	3	1 1/4	---	---	---	---	---	---			
120	100	400	3	---	8	402	397	373	229	---	1 1/4	1 1/4	1 1/4
120	100	400	1 AL	---	6 AL	---	---	---	---	222	2	2	2
120	200	800	3/0	2	---	390	388	363	---	---			
120	200	800	3/0	---	6	---	---	---	201	---	Not Made	2 1/2	2 1/2
120	200	800	250 AL	---	4 AL	---	---	---	---	195	Not Made	3	3

(1) Applies only when steel conduit is used.
(2) Per NEC Table 250-95. Applicable to non-metallic conduit runs.
(3) Based on: 3 Ungrounded Circuit Conductors
1 Full Size Neutral Conductor
1 Equipment Gr. Conductor Sized Per Table 250-95 Minimum
(Assumed Bare Copper/Insulated Alum.)
(4) NEC Wire Fill Table Permits Smaller Conduit Size

Note: Software is not limited to above examples

Figure 2–3A Table showing examples of acceptable equipment-grounding conductor lengths using steel conduit and tubing as the equipment grounding conductor on 120 volt to ground circuits.

Examples of Maximum Length Equipment Grounding Conductor (Steel EMT, IMC, GRC, Copper, or Aluminum)
Computed As A Safe Return Fault Path To Overcurrent Device
Based on 1994 Georgia Tech Software (Version 1.3)
With an Arc Voltage of 40 and 4 IP at 25° C Ambient

Circuit Voltage To Ground	Overcurrent Device Rating Amperes (75° C)	400% (4IP) Overcurrent Device Rating Amperes	Circuit Conductor Size AWG-kcmil Copper or Aluminum	(1) Conduit Trade Size	(2) Equipment Grounding Conductor Size Copper or Aluminum	Length of EMT Run Computed Maximum (In Feet)	Length of IMC Run Computed Maximum (In Feet)	Length of GRC Run Computed Maximum (In Feet)	(2) Copper Grounding Conductor Max Run (In Feet)	(2) Aluminum Grounding Conductor Max Run (In Feet)	(3) REQUIRED CONDUIT SIZE BASED ON NEC 310-16/250-95 & TABLES 1, 4 AND 5 OF CHAPTER 9		
											ENT	SCH 40 PVC	SCH 80 PVC
277	20	80	12	1/2	---	1170	1179	1140	---	---			
277	20	80	12	---	12	---	---	---	890	---	3/4	1/2	3/4
277	20	80	10 AL	---	10 AL	---	---	---	---	870	3/4	3/4	3/4
277	30	120	10	1/2	---	1199	(4) 1135	1143	---	---			
277	30	120	10	3/4	10	---	1182	---	946	---	3/4	3/4	3/4
277	30	120	10	---	10	---	---	---	---	920			
277	30	120	8 AL	3/4	8 AL	---	(4) 1228	---	---	---			1
277	40	160	8	1	---	1197	1276	1239	---	---			
277	40	160	8	---	10	---	1186	1131	---	---	1	1	1
277	40	160	8	---	10	---	---	---	676	---			
277	40	160	8 AL	---	8 AL	---	---	---	---	690	1	1	1 1/4
277	60	240	6	1	---	1192	1176	1107	---	---			
277	60	240	6	---	10	---	---	---	680	---	1	1	1 1/4
277	60	240	4 AL	---	6 AL	---	---	---	---	657	1 1/4	1 1/4	1 1/4
277	100	400	3	1 1/4	---	1157	1155	1077	---	---			
277	100	400	3	---	8	---	---	---	680	---	1 1/4	1 1/4	1 1/2
277	100	400	1 AL	---	6 AL	---	---	---	---	659	2	2	2
277	200	800	3/0	2	---	---	---	---	---	---			
277	200	800	3/0	---	6	---	---	---	598	---	Not Made	2 1/2	2 1/2
277	200	800	250 AL	---	4 AL	---	---	---	---	578	Not Made	3	3

(1) Applies only when steel conduit is used.
(2) Per NEC Table 250-95. Applicable to non-metallic conduit runs.
(3) Based on: 3 Ungrounded Circuit Conductors
1 Full Size Neutral Conductor
1 Equipment Gr. Conductor Sized Per Table 250-95 minimum
(Assumed Bare Copper/Insulated Alum.)
(4) NEC Wire Fill Table Permits Smaller Conduit Size

Note: Software is not limited to above examples

Figure 2–3B Table showing examples of acceptable equipment-grounding conductor lengths using steel conduit and tubing as the equipment grounding conductor on 277 volt to ground circuits.

Section 110–16 & 33 should also provide additional room for future expansion. Spare raceway capacity simplifies rewiring and is a significant factor in minimizing the cost when such work is undertaken.

> **Note:** The *NEC*® does not require the additional expansion capabilities recommended in the section.
>
> May 1994 Equipment Grounding Path Studies
> Georgia Tech Research Institute
> Dr A.P. Sakis Melipolous & Elias N. Glytsis
>
> The information in Figures 2–3A and 2–3B has been formulated from the computer software developed by Dr. Meliopolous, based on modeling and actual field tests by Kearney Laboratories, McCook, IL. These tests coincide with extensive changes to the 1996 *NEC*® which added exact dimension tables for each type of wire and raceway to Chapter 9 Tables 4 and 5 and a new Appendix C for fill capacity. An additional change added a Note 5 to Table 250–95 to warn the user that the minimum-size equipment-grounding conductor listed in that table may not be large enough in some cases. Figures 2–3A and 2–3B show several common examples of acceptable circuit lengths and conduit and tubing types.

FLEXIBILITY

A good electrical system provides flexibility. The distribution system should be arranged so as to accommodate future physical changes in plant or office layout as production lines are changed, equipment shifted, or offices rearranged. The increased popularity of modular office layouts demands an electrical system flexibility to meet quick and economical rearrangement. One or more raceway types will offer the versatility and flexibility needed to meet today's constantly changing world (see (FPN) following the definition of "Raceway" in Article 100 for a list of raceway types).

Steel tubular conduit (Rigid Metal Article 346, IMC Article 345, and EMT Article 348) offers almost unlimited expansion possibilities and is adaptable to all types of building construction and floor plans. In addition, cellular and under-floor raceways offer creative circuit and outlet arrangement. Similarly surface raceway, busway, and wireways provide additional flexibility in the design of industrial, institutional, and commercial facilities.

Flexibility can only be maintained if adequate additional space is provided in the panelboards and switchboards (see dedicated space requirements *NEC*® Section 384–4). Additional space should be provided, if possible, to add feeders in the future, or spare conduits can be installed to areas that either are not accessible or where future installation would be labor intensive such as under floor, buried in concrete, or in shafts. The importance of flexibility cannot be overemphasized.

ELECTRICAL SYSTEMS AND BUILDING FIRES

The prime objectives of the *National Electrical Code*® are to provide safety from fire and electrical shock (see Figure 2–4). Many of the Code's requirements are based on the premise that an electrical system designed and installed in accordance with the *National Electrical Code*® must not start a fire. In addition, an electrical installation must not contribute to or allow the spread of fire should one originate from other causes. The possible contribution of an electrical system to a building fire is undergoing more and more scrutiny as these systems become more extensive and as more uses are proposed for an increasing variety of electrical raceway materials.

Fire Resistance and Spread of Fire and Smoke

NFPA 101, Fire and Life Safety Code, and the three major model building codes (ICBO, BOCA, and

Figure 2–4 Example of fire damage caused by an electrical system failure. (Courtesy of Allied Tube and Conduit)

SBCCI), plus a variety of local, state, and city building codes, include many provisions whose purpose is to provide building construction that will not only withstand the effects of fire in a portion of a building, but will also reduce the likelihood of fire spreading within the building. For extremely large or high structures, building codes usually require fire-resistive construction, as well as the use of non-combustible construction materials. Building codes prescribe that for fire-resistive buildings, the columns, beams, girders, walls, and floor constructions have a resistance to fire that is measured by a standard fire test procedure (ASTM E119–Standard Method of Fire Test for Building Construction and Materials).

The fire resistance of a structural assembly is determined from end point criteria included in this fire test standard. For load-bearing members, such as columns, beams and girders, maintenance of the load-bearing capability during and after fire exposure is a prime criterion. The temperature rise on the unexposed surface of an assembly is critical for walls and floors aimed at preventing ignition of combustibles and the spread of fire to the opposite side of such wall or floor assemblies. It is extremely important in fire-resistive construction that all building components adhere to approved design of fire-rated construction. Furthermore, during construction every effort should be made to avoid damage to the fire-resistive integrity of the construction by removing fire-protective coatings or coverings or by piercing through entire fire-rated assemblies.

REVIEW QUESTIONS

(Answers to some of these questions may be found in the introduction and 1996 *NEC®*.)

1. In which article and section of the *NEC®* are the requirements found for the protection against the spread of fire?
   Answer: Reference:

2. What does the term "nominal voltage" mean?
   Answer: Reference:

3. Name three wiring methods which provide "physical protection" to the enclosed conductors?
   Answer: Reference:

4. What is the minimum burial depth for rigid nonmetallic conduit in earth?
   Answer: Reference:

5. A box is generally required for making splices and taps. When are conductors allowed to be spliced in direct contact with earth without enclosing them in a junction box?
   Answer: Reference:

6. Metal raceways, boxes, cabinets, etc., in wet locations must be installed so as to provide an air space of at least _____ inches between them and the wall.
   Answer: Reference:

7. All raceways installed in direct contact with earth shall be sealed at both ends to prevent moisture from entering.
   Answer: True or False.   Reference:

8. All interior raceways exposed to widely different temperatures are required to be sealed.
   Answer: True or False.   Reference:

9. May cable tray systems be partitioned to enclose both electrical wires and pneumatic tubing for process controls?
   Answer: Reference:

10. What is the standard fire test method for building materials?
    Answer: Reference:

Chapter 3

Hazardous Locations

The engineers and contractors of today, handle the design and installation of electrical systems with a high level of confidence. The subject of hazardous (classified) locations, however, is still viewed with uncertainty by many people in the industry.

Actually, hazardous location installations are more exacting than difficult or mysterious. Article 500 of the *National Electrical Code®* defines various types of hazardous locations and Articles 501 through 517 provides hazardous performance and installation requirements.

There are many rules to guide you; however, assurance of safety starts with a good understanding of two basics: proper classification of locations and selection of correct equipment.

When an electrical system, or any portion of it, is to be installed in an area containing a hazardous atmosphere, protection against possible explosions becomes a major design factor. All of the basic design considerations still apply to determine conductor sizes and general layout of the overall system. In addition, the portion of the system in or passing through an area where flammable gases, combustible dust, or ignitable fibers may exist must have equipment enclosures especially designed to prevent electrical sparks and arcs from igniting the surrounding atmospheres and causing serious fires and/or explosions. For this portion of the design, the electrical engineer must be careful to select equipment specifically approved for the conditions encountered (see Figures 3–1, 3–2, and 3–3).

Figure 3–1 Electrical panelboard suitable for Class II & III Groups E, F, & G. (Courtesy of Crouse-Hinds, Division of Cooper Industries)

Figure 3–2 Electrical panelboard suitable for Class I, II, & III Groups C, D, E, F, & G. (Courtesy of Crouse-Hinds, Division of Cooper Industries)

> **Warning:** *NEC®* Section 90–3 states that Chapter 5 only modifies and augments Chapters 1 through 4. Corrosion, capacity, suitability, and service are not considered. Therefore, it is the responsibility of the designer to take these factors into account, along with the hazardous location requirements which are covered in Chapter 5.

The *National Electrical Code®* describes three classifications of hazardous locations where an uncon-

HAZARDOUS (CLASSIFIED) AREAS

Hazardous (classified) locations are classified based on the properties of flammable gases, vapors, liquids, combustible dusts, fibers and flyings which may be present and the likelihood that a concentration of flammable or combustible materials is present in sufficient quantity. Areas or sections of each building or room may be classified individually by indicating: the class, division, and group present.

Figure 3–3

CHAPTER 5 *NEC®* CLASSES-DIVISIONS-GROUPS

	Class I	
Division 1		Division 2
Group A		Group A
B		B
C		C
D		D
	Class II	
Division 1		Division 2
Group E		Group E
F		F
G		G
	Class III	
Division 1		Division 2
No Groups		No Groups

Figure 3–4

fined spark could result in explosion or fire (see Figure 3–4). Each of these classifications identifies conditions that may exist either under normal operations, under occasional, but frequent conditions resulting from maintenance operations, or under conditions of accidental breakdown that may cause the release of flammable or readily combustible materials. *NEC®* Sections 500 through 505, 510, 511, and 513 through 517 contain rules covering the type of wiring and apparatus to be installed in hazardous locations.

The actual degree of hazard (classification) is frequently evaluated by local inspection and fire department authorities, and for that reason, they should be consulted during the design stage of the hazardous area wiring system.

"Big Blasts"

Some of this nation's major accidents that have resulted in severe damage to property and loss of life have occurred in the chemical petroleum industries, food processing industries, and other locations containing hazardous areas. These facilities have qualified maintenance personnel and engineering, and they have taken precautions to avoid such accidents. However, the following examples emphasize the dangers in dealing with hazardous classified materials.

- April 8, 1992: A natural gas pipeline explodes near Brenham, Texas, killing one.
- September 3, 1991: A chicken-processing plant of Imperial Food Products Co. near Hamlet, North Carolina, catches fire, killing twenty-five.
- June 17, 1991: Tenneco Inc.'s phosphorus chemicals plant in Charleston, South Carolina, explodes, killing six.
- May 1, 1992: IMC Fertilizer Angus Chemical Co. nitro-paraffins plant in Sterling, Louisiana, explodes, killing eight.
- March 12, 1991: Union Carbide petrochemical plant in Seadrift, Texas, explodes, killing one.
- March 3, 1991: Citgo Petroleum Corporation petroleum refinery near Lake Charles, Louisiana, explodes, killing six.
- July 19, 1990: BASF Corporation in Cincinnati, Ohio, chemical plant explodes, killing two.
- July 5, 1990: Arco Chemical Co. petrochemical plant in Channelview, Texas, explodes, killing seventeen.
- October 23, 1989: Phillips Petroleum Co. plastics plant in Houston, Texas, explodes, killing twenty-three.
- December 24, 1989: Exxon Company petroleum refinery in Baton Rouge, Louisiana, explodes, killing two.
- May 4, 1988: Pacific Engineering and Production Company ammonium prechlorate plant in Henderson, Nevada, explodes, killing two.
- November 5, 1985: Warren Petroleum plant at Warren Belvieu, Texas, explodes, killing two.
- July 1984: Unocal Corporation petroleum refinery in Lemont, Illinois, explodes, killing seventeen.

NEC® CLASSIFICATIONS

The Code divides hazardous locations into three classes, plus divisions and groups which are defined as follows. These are based on the degree of hazard and the likelihood of this hazard being present. The general information regarding these areas are covered in Article 500. Each class is specifically covered by a separate article, i.e., Class I Article 501, Class II Article 502, and Class III Article 503. Some common hazardous installations are specifically covered by an article, i.e., Article 514 "Gasoline Dispensing, Service Stations."

Important fact: Article 514 covers specific requirements for gasoline-dispensing service stations; one must also apply pertinent sections of Article 502 and 500 and Chapters 1 through 4 to make the installations.

Classes

The NFPA standards divide the hazardous properties into three classes so that the installation containing one or more of those properties can be maintained and operated safely using the equipment and wiring methods that will operate safely with a minimum of over design.

Class I

This classification applies to locations where the hazard of fire or explosion may exist due to the presence of flammable liquids, gases, or vapors in quantities sufficient to produce an explosive or ignitable environment.

Warning: Class III locations are not necessarily less hazardous than Class I locations. Major life and property losses have occurred in Class I, Class II, and Class III locations.

Class II

This classification applies to locations where the hazard of fire or explosion may exist due to the presence of combustible dust.

Class III

This classification applies to location where the hazard of fire or explosion may exist due to the presence of easily ignitable fibers or flyings.

Divisions

The classification applies to each of the three classes and divides them into two divisions according to the degree of hazard based on the conditioning and handling of the substance.

Division I

This classification identifies the conditions of the area where the hazardous substance is "normally" expected to be present during day-to-day operations.

> One example of a Class II Division 1 area is an ammunition process line where explosive gun powder is used.

Division 2

This classification identifies the conditions of an area where hazardous location substances are *not* normally present in the atmosphere, except under abnormal conditions.

> One example of a Class I Division 2 area is when flammable liquids or gases are confined in closed containers or systems where they can't escape except through an accident.

Groups

> For the hazardous group atmosphere identifications, see *NEC*® Section 500–3(a) and (b).

Class I Groups A, B, C, and D are materials divided by flame propagation characteristics (the ability of the fire to multiply), explosion pressures (the force of an explosion if one should occur), and other factors related to the substance that is present or may be present.

> For example: Group A is an atmosphere containing acetylene.

Class II Groups E, F, and G define the ignition temperature (the temperature that will cause spontaneous combustion or self-ignition) and conductivity of each of the types of combustible dusts.

> For example: Group E is atmospheres containing aluminum, magnesium, their alloys, and other similar combustible dusts.

Class III materials are not classified in groups.

EVALUATION OF HAZARDOUS AREAS

Each area that contains gases or dusts that are considered hazardous must be carefully evaluated to make certain the correct electrical equipment is selected. Many hazardous atmospheres are Class I Group D, or Class II Group G. However, certain areas may involve other groups, particularly Class I Groups B and C. Conformity with the *National Electrical Code*® requires the use of enclosures and fittings approved for the specific hazardous gas or dust involved. In some instances, Group D equipment can be used where Group B atmospheres can be sealed off (see Sec 500–3(a)(2) Exception 1).

CLASSIFICATION OF HAZARDOUS AREAS

Classification of specific areas is determined by the AHJ "code enforcing authority"—who may be a representative of the insurance underwriters, a municipal electrical inspector, a fire marshal, or a member of the corporate safety organization.

Even though the professionals who classify locations "live by the Code," many factors may not be obvious. The distance from the hazardous source may affect the classification of the area. Ventilation and air currents must be considered. Temperatures, topography, walls, and special safeguards may also have an effect.

Incorrect classification could lead to using equipment that may not be safe. This may result in fires, explosions, and even loss of life and property. With so many code requirements and location conditions to evaluate, experience, technical knowledge, and judgment are essential for classification of hazardous areas.

Hazardous locations are those areas where a potential for explosion and fire exist because of flammable gases, vapors, or finely pulverized dusts in the atmosphere because of the presence of easily ignitable fibers or flyings. Hazardous locations may result from the normal processing of certain volatile chemicals, gases, grains, etc., or it may result from accidental failure of storage systems for these materials. It is also possible that a hazardous location may be created when volatile solvents or fluids, used in a normal maintenance routine, vaporize to form an explosive atmosphere.

Regardless of the cause of a hazardous location, it is necessary that every precaution be taken to guard against ignition of the atmosphere. Certainly no open flames would be permitted in these locations, but what about other sources of ignition?

> Example: I inspected an ammunition production line where the likelihood of an incident was so high that the designer had the production line celled off with 24-inch concrete walls every 10 to 12 feet. This design would limit the loss and damage to a small area if an explosion occurred. He had also provided many other innovative ideas to minimize dangers in this extremely high-hazard location.

Electrical Sources of Ignition

A source of ignition is simply the energy required to touch off an explosion in a hazardous location atmosphere (see Figure 3–5). When these three elements—oxygen, fuel, and ignition—are present, a fire or explosion will occur. The quantity and volatility of the fuel and oxygen present (fuel–oxygen mix) will determine the extent of the hazard present. The degree of heat required to ignite that mixture depends upon the type of fuel present.

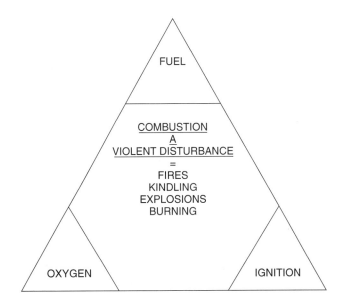

Figure 3–5 These three ingredients combined are needed to have a fire.

> Example: The flash point for gasoline is –36° to –50°F, as compared to the flash point for kerosene which is 110° to 162°F. This comparison shows that it may take an open flame to ignite kerosene and only a spark to ignite gasoline.

In the example above, one can see that generally fuels requiring an open flame to ignite may be wired in accordance with the first four chapters of the code and are not considered hazardous (fuels with a flash point above 100°F). Where more hazardous substances are present, it is necessary to consider Chapter 5 methods. However, only when all of the factors are known to the designers can they make that final determination for requiring Chapter 5 wiring methods and equipment suitable for hazardous (classified) locations which will minimize the dangers in the day-to-day operations.

The designer must recognize that the electrical equipment can also be a source of this ignition energy. The normal operation of switches, circuit breakers, motor starters, contactors, plugs, and receptacles release this energy in the form of arcs and sparks as contacts open and close, making and breaking circuits. Electrical equipment such as lighting fixtures and motors are classified as "heat producing," and they will become a source of ignition if they reach a surface temperature which exceeds the ignition temperature of the particular gas, vapor, or dust in the atmosphere.

It is also possible that an abnormality or failure in an electrical system may provide a source of ignition. A loose termination in a splice box or a loose lamp in a socket can be the ignition source of both. The failure of insulation from cuts, nicks, or aging can also act as an ignition source from sparking, arcing, and heat.

Look for the Marking or Label

The *NEC*® contains rules for selection and installation of electrical equipment. Section 500–2(c) and (d) requires equipment for hazardous (classified) locations to be clearly approved and marked with the class, group, and temperature for which it is designed (see Figures 3–5A, 3–5B, and 3–6). The code states that the temperature will be marked as a maximum safe operating temperature or as an identification

(d) Marking. Approved equipment shall be marked to show the class, group, and operating temperature or temperature range referenced to a 40°C ambient.

(FPN): Equipment not marked to indicate a division, or marked "Division 1" or "Div. 1," is suitable for both Division 1 and 2 locations. Equipment marked "Division 2" or "Div. 2" is suitable for Division 2 locations only.

The temperature range, if provided, shall be indicated in identification numbers, as shown in Table 500-3(d).

Exception: As required in Section 505-10(b).

Identification numbers marked on equipment nameplates shall be in accordance with Table 500-3(d).

Exception: As required in Section 505-10(b).

Equipment that is approved for Class I and Class II shall be marked with the maximum safe operating temperature, as determined by simultaneous exposure to the combinations of Class I and Class II conditions.

Exception No. 1: Equipment of the nonheat-producing type, such as junction boxes, conduit, and fittings, and equipment of the heat-producing type having a maximum temperature not more than 100°C (212°F) shall not be required to have a marked operating temperature or temperature range.

Exception No. 2: Fixed lighting fixtures marked for use in Class I, Division 2 or Class II, Division 2 locations only shall not be required to be marked to indicate the group.

Exception No. 3: Fixed general-purpose equipment in Class I locations, other than fixed lighting fixtures, that is acceptable for use in Class I, Division 2 locations shall not be required to be marked with the class, group, division, or operating temperature.

Exception No. 4: Fixed dusttight equipment other than fixed lighting fixtures that are acceptable for use in Class II, Division 2 and Class III locations shall not be required to be marked with the class, group, division, or operating temperature.

Exception No. 5: Electric equipment suitable for ambient temperatures exceeding 40°C shall be marked with both the maximum ambient temperature and the operating temperature or temperature range at that ambient temperature.

Table 500-3(d). Identification Numbers

Maximum Temperature Degrees C	Degrees F	Identification Number
450	842	T1
300	572	T2
280	536	T2A
260	500	T2B
230	446	T2C
215	419	T2D
200	392	T3
180	356	T3A
165	329	T3B
160	320	T3C
135	275	T4
120	248	T4A
100	212	T5
85	185	T6

(FPN): Since there is no consistent relationship between explosion properties and ignition temperature, the two are independent requirements.

Figure 3–5A Sections 500–3(d) and Table 500–3(d). (Reprinted with permission from NFPA 70–1996, *National Electrical Code*®, Copyright © 1995, National Fire Protection Association, Quincy, Massachusetts 02269. This reprinted material is not the complete and official position of the NFPA on the referenced subject, which is represented only by the standard in its entirety)

(e) Class I Temperature. The temperature marking specified in (d) above shall not exceed the ignition temperature of the specific gas or vapor to be encountered.

Exception: Where area classification is in accordance with Article 505, the temperature marking in Section 505-10(b) shall not exceed the ignition temperature of the specific gas or vapor to be encountered.

(FPN): For information regarding ignition temperatures of gases and vapors, see *Manual for Classification of Gases, Vapors, and Dusts for Electrical Equipment in Hazardous (Classified) Locations*, NFPA 497M-1991, and *Guide to Fire Hazard Properties of Flammable Liquids, Gases, and Volatile Solids*, NFPA 325-1994.

(f) Class II Temperature. The temperature marking specified in (d) above shall be less than the ignition temperature of the specific dust to be encountered. For organic dusts that may dehydrate or carbonize, the temperature marking shall not exceed the lower of either the ignition temperature or 165°C (329°F). See *Manual for Classification of Gases, Vapors, and Dusts for Electrical Equipment in Hazardous (Classified) Locations*, NFPA 497M-1991, for minimum ignition temperatures of specific dusts.

The ignition temperature for which equipment was approved prior to this requirement shall be assumed to be as shown in Table 500-3(f).

Table 500-3(f)

Class II Group	Equipment That Is Not Subject to Overloading		Equipment (Such as Motors or Power Transformers) That May Be Overloaded			
			Normal Operation		Abnormal Operation	
	Degrees C	Degrees F	Degrees C	Degrees F	Degrees C	Degrees F
E	200	392	200	392	200	392
F	200	392	150	302	200	392
G	165	329	120	248	165	329

Figure 3–5B Section 500–3(e), (f) and Table 500–3(f). (Reprinted with permission from NFPA 70–1996, *National Electrical Code*®, Copyright © 1995, National Fire Protection Association, Quincy, Massachusetts 02269. This reprinted material is not the complete and official position of the NFPA on the referenced subject, which is represented only by the standard in its entirety)

number, which is listed in Table 500–3(d) of the Code. There are five exceptions to the general rule.

In most cases there will be a trademark of the testing organization on the label such as Underwriters Laboratory (UL), Factory Mutual (FM), or Canadian Standards Association (CSA). These organizations are certifying agencies. Their trademark means that the equipment has been properly tested and has met the standards for the application specified. However, it should be noted that the NFPA® committee responsible for Chapter 5 CMP–14 has specified approved equipment rather than listed equipment in many cases. They have substantiated this by stating that much equipment designed and manufactured for hazardous (classified) locations is not mass-produced, and therefore, listing would substantially increase costs and delay delivery. (Review definitions of "approved" and "listed" in Article 100; many inspectors will require the equipment to be listed before they will approve it.)

Figure 3–6 Examples of labels as they appear on equipment suitable for hazardous locations. (Courtesy of Underwriters Laboratories)

HAZARDOUS LOCATIONS AND THE *NATIONAL ELECTRICAL CODE*®

The *National Electrical Code*® treats installations in hazardous locations in Articles 500 through 517. Each hazardous location can be classified by the definitions in the *NEC*®, and following are interpretations of these classifications and applications.

EXPLOSION-PROOF APPARATUS

An explosion-proof apparatus is equipment which is contained in an approved enclosure that is marked and labeled for the application and is capable of withstanding (containing) an explosion of a specified gas or vapor which may occur within it and of preventing the ignition of a specified gas or vapor surrounding the enclosure by sparks, flashes, or explosion of the gas or vapor which originate within the enclosure. It is an apparatus and enclosure which operates in a manner so that the external temperature of the enclosure that will not reach temperatures that will ignite the surrounding flammable atmosphere (see Figure 3–6A and Figure 3–7).

Figure 3–7 Explosion-proof combination switch and receptacle. (Courtesy of Crouse-Hinds, Division of Cooper Industries)

Figure 3–6A Class I Switch Rack. This shows the methods acceptable for this hazardous location. (Courtesy of Crouse-Hinds, Division of Cooper Industries)

DUST-IGNITION PROOF

NEC® 502–1 states that "as used in this Article" shall mean enclosed in a manner that will exclude ignitable amounts of dust that might affect performance or rating and that, when installed and protected in accordance with this code, will not permit arcs, sparks, or heat otherwise generated or liberated inside of the enclosure to cause ignition of exterior accumulations or atmospheric suspensions of a specified dust on or in the vicinity of the enclosure. Class II equipment is designed to prevent the entrance of dust particles. However, the accumulation of ignitable dust on the enclosure itself is not controllable; therefore, the term "dust-ignition proof" means that the enclosure will not reach temperatures high enough for the dust which accumulates on these enclosures to ignite.

HAZARDOUS LOCATION

This is an area where the possibility of explosion and fire is created by the presence of flammable gases, vapors, dusts, fibers, or flyings. These are areas which meet the definition of one of the three classes, Class I, II, or III as described below.

Class I

Class I is composed of those areas in which flammable gases or vapors may be present in the air in sufficient quantities to be explosive or ignitable (see Appendix A–1a and b and Figure 3–8). These are areas or locations in which the materials are used in production or in day-to-day operations, or they may be areas in which the materials are simply stored and not handled at all.

Class II

Class II is composed of those areas made hazardous by the presence of combustible dust (see Appendix A–2

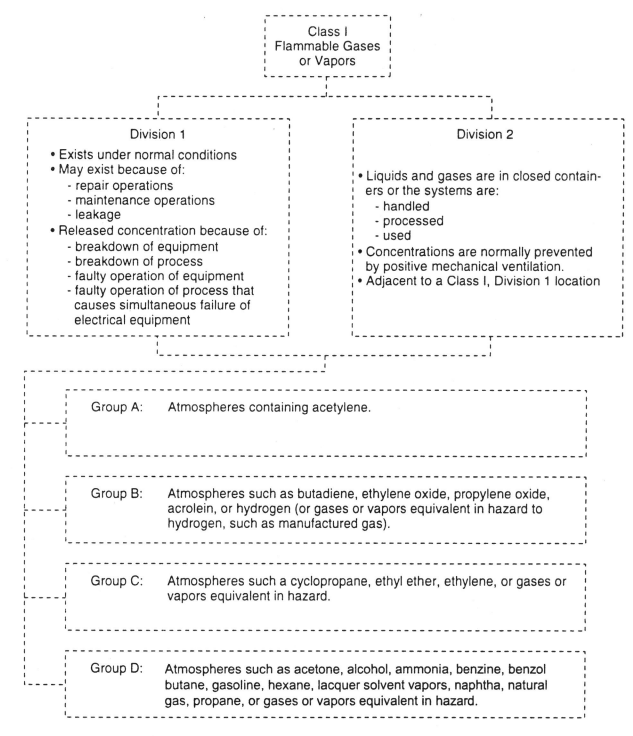

Figure 3–8 Class I Classification by Class, Division, and Group in accordance with Article 500.

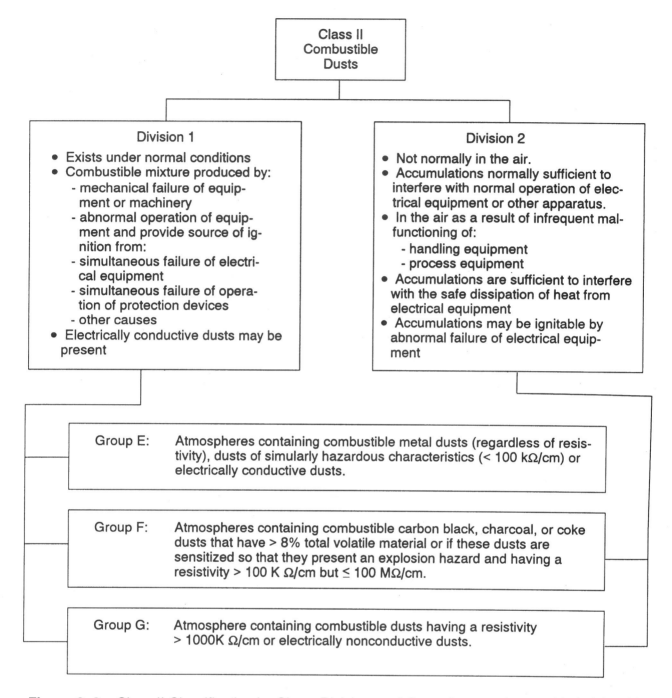

Figure 3–9 Class II Classification by Class, Division, and Group in accordance with Article 500.

and Figure 3–9). These are areas or locations similar to the Class I locations in which the materials (combustible dusts) are used in production or in day-to-day operations, or they may be areas in which the materials are simply stored and not handled at all.

Class III

Class III is composed of those areas in which there are easily ignitable fibers or flyings present, due to the type of material being handled, stored, or processed (see Figure 3–10). Many believe that the classes are degrees of hazards, which is not true. Each class can be equally as hazardous as the other. Class III has a very high rate of incidence. These areas are typically cabinet shops, furniture factories, textile mills, etc. We often think of these areas as nonhazardous, but they can be very hazardous if they are not kept very clean. All equipment must be cleaned regularly so that accumulation and build up of particles and

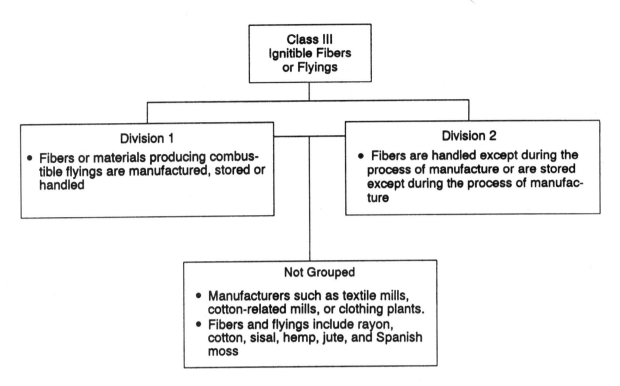

Figure 3–10 Class III Classification by Class and Division in accordance with Article 500.

flyings are controlled. Many of the materials that are Class III materials will ignite from the surface temperature of equipment that is in the area but is not specifically approved for the application.

Fibers and flyings are small particles of cloth or wood and are often natural and synthetic fibers.

Division 1 in the Normal Situation

This hazard would be expected to be present in everyday production operations or during frequent repair and maintenance activity.

Division 2 in the Abnormal Situation

The material is expected to be confined within closed containers or closed systems and will be present only through accidental rupture, breakage, or unusual faulty operation.

Groups

The gases and vapors of Class I locations are broken into four groups by the code: A, B, C, and D. These materials are grouped according to their explosion pressures and other flammable characteristics. Class II locations are grouped as E, F, and G. These groups are classified according to the *ignition temperature* and the *electrical conductivity* of the hazardous substance.

Seals

Special fittings are required in Class I locations to minimize the passage of gases and vapors or, in the case of an explosion, prevent the passage of flames (see Figures 3–11, 3–12, and 3–13). They are also required in Class II locations to prevent the passage of combustible dust.

Article 500

Article 500 through 503 (1996 *NEC*®) explains in detail the requirements for the installation of wiring or electrical equipment in hazardous locations. As previously stated in this text, these articles along with other applicable regulations, local governing inspection authorities, insurance representatives, and qualified engineering/technical assistance should be your guides to the installation of wiring or electrical equipment in any hazardous or potentially hazardous location.

Figure 3–11 Example of an explosion-proof seal fitting suitable for horizontal or vertical installation. (Courtesy of Crouse-Hinds, Division of Cooper Industries)

CLASSIFICATION OF AREA

To classify an area requires knowledge of the basic application intended for the installation to be classified. It is important to have a complete understanding of the equipment which is needed and required for the day-to-day operations of that area. It is also important to rely on experience and sound judgment. Electrical area classification is performed to prevent explosions and fires caused by electrical equipment arcing and sparking, high temperatures, or electrical system faults, and to prevent personal injury,

Figure 3–12 Cutaway view of an installed vertical seal fitting. (Courtesy of Crouse-Hinds, Division of Cooper Industries)

Figure 3–13 Cutaway view of an installed horizontal seal fitting. Two fiber dams are required in this application. (Courtesy of Crouse-Hinds, Division of Cooper Industries)

loss of life, and destruction of buildings and equipment. The installation of correct electrical equipment and materials and the use of the correct installation techniques can make a major contribution towards significantly reducing the risk of such occurrences.

> No one individual should be totally responsible for area classification!

No one individual should be totally responsible for the ultimate classification or the decision for classifying an area. It should be the result of group meetings and agreement with the customer who can provide experience and the history related to the proposed area or to a similar area using similar hazardous materials. Experienced plant personnel, plant operations, plant maintenance, plant safety department, and in-house engineering should be involved and consulted. Electrical engineers involved with the design, architectural designers involved with the architectural appurtenances, the authority having jurisdiction, which may consist of local and state electrical, fire inspectors, and building officials, and the operating company's underwriters (insurers) should also be consulted. Careful coordination among these groups will not only ensure the correct classification, but will also provide alternatives to reduce the hazardous areas to a minimum. Many industry codes and standards are available. (**Note:** There may also be local codes which need to be consulted.) They will provide valuable guidance for those ultimately responsible for making these decisions.

> NFPA-70, the *National Electrical Code*® contains the *minimum mandatory* electrical installation requirements.

You must first have an industry code that is uniquely related to the industry, chemical or hazardous material for which this area will be designed, such as API (the American Petroleum Institute). They have unique recommendations for the classification of areas dealing with petroleum facilities, such as API IRP 500-1992. Other industries have similar application codes or guides available. The representative for the operating company should be able to provide this information. There are many standards available through the National Fire Protection Association which should assist in making this determination.

- NFPA 497 concerns the classification of Class I hazardous locations for electrical installations in chemical plants.
- NFPA 497-A-1992 covers Class I hazardous locations in chemical process areas.
- NFPA 497-B-1991 is a classification of Class II hazardous locations for electrical installations in chemical process areas. NFPA 497 is a useful document and may be used in conjunction with other industry standards; however, there may appear to be conflicting requirements in some instances.
- NFPA 497-M-1991 is a classification for gases, vapors, and dusts for electrical equipment in hazardous locations.
- NFPA 70, the *National Electrical Code*®, Chapter 5 are mandatory requirements as adopted by the jurisdiction in which the facility is being built. You must check with the authority having jurisdiction in that area, state, county, or city, to verify which edition of NFPA-70 *National Electrical Code*® has been adopted and if there are any amendments.
- NFPA-321 covers the classification of flammable and combustible liquids.
- NFPA-325-M-1987 describes the fire hazard for those flammable liquids, gases, and volatile solids. This provides additional information to help determine the gas groups that are not described in NFPA-497-M or elsewhere.
- NFPA-496-1993 deals with purged and pressurized enclosures.

> **Warning:** NFPA 496 is only applicable to Class I installations. Purging and pressurizing of enclosures is not feasible for Class II applications.

- NFPA-493 deals with intrinsically safe apparatus and associated apparatus for use in Class I, II, and III Division 1 hazardous locations.
- NFPA-30-1987 is useful for flammable and combustible liquids and covers the ventilation, handling, and storage of such combustible liquids.

- Other standards are available and may be useful in making the proper classification.

After the areas are clearly defined and the class has been established, the group then must also be established. The group is listed in alpha designations, A through D for Class I locations, and E through G for Class II locations. Class III locations are not divided into groups. NFPA–70, gives a partial list of the most commonly encountered materials which have been tested for grouping. They are grouped on the basis of their flammability characteristics, Group A through G, with Group A being the most volatile, acetylene. Groups E through G are flammable volatile dusts. The responsible party, such as the designer or contractor, should utilize the expertise of chemical process engineers, safety loss prevention engineers, or other experts in developing and establishing the group and class for the area being designed.

Next, the division must be established. NFPA–70, *National Electrical Code®*, recognizes two divisions. Division criteria is based upon the presence of volatile liquids, vapors, or dust. A Division 1 location is one that is likely to have volatile liquids, vapors, or dust present under normal conditions, such as an area where the volatile material is dispensed. A Division 2 location is an area that is likely to have flammable or volatile gases, vapors, or dust present only under abnormal conditions, such as accidental spills or accidental releases of the volatile material present.

One should classify the division very carefully. An incorrect area classification of Division 2 permits a much more liberal wiring method with fewer safety guards built in. Improper classification of an area can greatly reduce the safety factor desired under all operating conditions. It is possible to reduce the classification but only if assurance is made that all materials are maintained through ventilation and there is good area janitorial maintenance. However, operation management personnel often plan on good maintenance and a clean facility to meet the lower classification requirements, but within a few months or years the hazards present within the facility due to poor janitorial maintenance are creating a much more hazardous condition than the facility was originally designed for. Existing conditions in a similar facility under the same ownership or management, if available, may provide valuable information to be considered during the classification process by those responsible.

Additional references for the application and use of hazardous materials may be found in Appendix A of NFPA–70 of the *National Electrical Code®*.

After all facts that are available are known and a classification is assigned to each building or area, the selection of the materials and the installation must be made in accordance with the adopted edition of *NEC®* and any local adopted amendments as a minimum installation standard. Many engineers will specify materials and methods more stringent than those in the *NEC®*.

REVIEW QUESTIONS

(Answers to some of these questions may be found in the introduction, previous chapters, and 1996 *NEC®*.)

1. In which code are the minimum installation requirements found?
   Answer: Reference:

2. Where butadiene is sealed in accordance with Section 501–5, the raceway system may be classed _____ Division _____ Group _____.
   Answer: Reference:

3. Gasoline, kerosene, and propane are Class I Group D.
   Answer: True or False.   Reference:

4. Who must "approve" hazardous equipment?
   Answer: Reference:

5. Why are motors and light fixtures required to be marked with their operating temperature?
   Answer: Reference:

6. An area where the hazardous substance is present under normal conditions is a Class _____ location.
   Answer: Reference:

7. A grain silo is a Class _____ hazardous location.
   Answer: Reference:

8. A paint spray booth is a Class _____ Division _____ hazardous location.
   Answer: Reference:

9. May a seal fitting be installed inside the hazardous (classified) area before the conduit leaves that area?
   Answer: Reference:

10. An area where combustible flyings are produced from a manufacturing process is considered a Class _____ Division _____ area.
    Answer: Reference:

Chapter 4

Class I Hazardous (Classified) Areas

CLASS I LOCATIONS

Class I locations as covered in *NEC®* Article 501 are those in which flammable gases or vapors are or may be present in the air in quantities sufficient to produce explosive or ignitable mixtures (see Figure 4–1).

Figure 4–1 Switch rack assembly and lighting in a Class I installation. (Courtesy of Crouse-Hinds, Division of Cooper Industries)

Class I Location Examples

- Petroleum refining facilities
- Dip tanks containing flammable or combustible liquids
- Dry cleaning plants
- Plants manufacturing organic coatings
- Spray finishing areas (residue must be considered)
- Petroleum dispensing areas
- Solvent extraction plants
- Plants manufacturing or using pyroxylin (nitrocellulose) type and other plastics (Class II also)
- Locations where volatile inhalation anesthetics are used and stored
- Utility gas plants, operations involving storage and handling of liquefied petroleum and natural gas
- Aircraft hangars and fuel servicing areas

Class I Group Classifications

Since the different vapors and gases making up hazardous atmospheres have varying properties, the properties are placed in groups based on common flame propagation characteristics and explosion pressures as defined in *NEC®* Section 500–32(a). These groups are designated A, B, C, and D. The equipment selected must be suitable for the group of the specific hazardous gas or vapor with regard to flame propagation, explosion pressures, and operating temperatures.

Reference to the *National Electrical Code®* will indicate that most of the equipment used for Class I Division 2 applications is the same as that used for Division 1 applications. However, in certain instances, standard location equipment may be used for some of the Class I Division 2 applications if the appropriate restrictions are followed. The *National Electrical Code®* contains the specific rules.

Class I
Flammable gases and vapors

Division 1
Condition exists under normal conditions
May exist during repairs or maintenance

Group A—Atmosphere containing acetylene

Group B—Atmospheres containing butadiene, ethylene oxide, propylene oxide, acrotein, or hydrogen (also includes gases of equivalent hazard)

Group C—Atmospheres containing cyclopropane, ethyl ether, ethylene, or gases or vapors equivalent in hazard

Group D—Atmospheres containing acetone, alcohol, ammonia, benzene, Bengal, butane, gasoline, hexane, lacquer solvent vapors, naphtha, natural gas, propane, or gases or vapors of equivalent hazard

---

CLASS I DIVISION 1

Class I Division 1 locations are locations where the hazardous material, flammable gases, or vapors are or may be present in the atmosphere during normal operations. Areas where, during maintenance, repair, breakdown, and faulty operation, the presence of ignitable concentrations of gases or vapors can or may be released either continuously, intermittently, or periodically are classified as Division 1 locations. Locations where a breakdown in the operation of processing equipment results in the release of hazardous vapors and the simultaneous failure of electrical equipment are also Class I Division 1 locations (see Figures 4–2, 4–3, and 4–4).

Examples of Class I Division 1 Locations

- Locations where volatile flammable liquids or liquefied flammable gases are transferred from one container to another
- Paint spray booths
- Areas where there are open vats or tanks of flammable liquids
- Drying rooms or compartments for the evaporation of flammable liquids
- Portions of dyeing or cleaning plants where flammable liquids are used
- Gas generation rooms
- Inadequately ventilated pump rooms for flammable liquids
- Interiors of refrigerators and freezers where flammable liquids are stored

Figure 4–2 Example of a Class I location EXGJF or EXLK flexible coupling. Ideal to use with motors and other equipment where flexibility is needed. (Courtesy of Appleton Electric Company)

Class I Hazardous (Classified) Areas 39

Figure 4–3 Power Diagram for a Class I Division 1 location. (Courtesy of Appleton Electric Company)

CLASS I, DIVISION 1

NEC® 501-1. GENERAL
The general rules of this Code shall apply to the installation of electrical wiring and equipment in locations classified as Class I under Section 500-5 except as modified by this Article.

NEC® 501-2. TRANSFORMERS AND CAPACITORS.
Not a product of Appleton.

NEC® 501-3. METERS, INSTRUMENTS AND RELAYS.
1—GUB/GUBB EXPLOSION-PROOF ENCLOSURES (Section K).
1A—GUBM/GUBBM EXPLOSION-PROOF ENCLOSURES WITH GLASS FOR VIEWING OF INSTRUMENTS (Section K).

NEC® 501-4. WIRING METHODS.
2—GRU/GRUE CAST CONDUIT OUTLET BOX UNILETS (Section J).
2A—GUB CAST JUNCTION BOX (Section K).
3—ELBD PULLING UNILET (Section J).
4—UNF and UNF (NR) UNIONS (Section I).
5—ELF ELBOWS, PLUGS AND MISCELLANEOUS FITTINGS (Section I).
6—EXLK FLEXIBLE COUPLINGS (Section I).

NEC® 501-5. SEALING.
7—EY SEALING UNILETS (Section I).
8—ECDB DRAIN/BREATHER (Section I).

NEC® 501-6. SWITCHES, CIRCUIT BREAKERS, MOTOR CONTROLLERS AND FUSES.
9—XLP/XLB or XPP NON-FACTORY SEALED PANELBOARDS (Section P); Sealing Unilets (Section I) required. Also refer to NEC® Art. 430.
9A—EPP FACTORY SEALED PANELBOARDS; Sealing Unilets (Section I) required. Also refer to NEC® Art. 430.
9B—EB CIRCUIT BREAKER UNILETS (Section O). Also refer to NEC® Art. 430.
10—ETC COMBINATION CIRCUIT BREAKER AND MOTOR STARTER UNILETS WITH OPTIONAL PUSHBUTTONS INSTALLED (Sections O & M). Also refer to 430.11—EFS SWITCH UNILETS. Also refer to NEC® 430 (Sections N & O).
11A—EDS FACTORY SEALED MOTOR STARTER Unilets® (Section O).
11B—INTRAGROUND™ NI SERIES MOTOR STARTER (Section O); sealing Unilets® required (Section I). Also refer to NEC® 430.
12-EFD FACTORY SEALED PUSH BUTTON CONTROL STATION UNILETS (Section M). Also refer to NEC® 430.

NEC® 501-7. CONTROL TRANSFORMERS AND RESISTORS.

NEC® 501-8. MOTORS AND GENERATORS.
Not a standard product of Appleton, but motors provided with some Reelite® products (Section W3).

NEC® 501-9. LIGHTING FIXTURES.
Best described on next page.

NEC® 501-10. UTILIZATION EQUIPMENT, FIXED AND PORTABLE.
UTILIZATION EQUIPMENT—Products such as portable lamps must be approved for Class I locations.
14—ECC FACTORY SEALED CORD CONNECTORS (Section L).
The Appleton ECC is an approved method for permanent connection from a rigid conduit system to a portable appliance. Supplies power to utilization equipment.

NEC® 501-11. FLEXIBLE CORDS, CLASS I, DIVISION 1 AND 2.
Not a standard product listing, but flexible cord is provided with Appleton Reelite® designed for hazardous locations.

NEC® 501-12. RECEPTACLES AND ATTACHMENT PLUGS, CLASS I, DIVISIONS 1 AND 2.
15—EFS U-LINE® FACTORY SEALED PLUGS AND RECEPTACLES (Section L); sealing Unilets® not required.
16—EBR PLUGS AND RECEPTACLES (Section L); sealing Unilets® required.

NEC® 501-13. CONDUCTOR INSULATION, CLASS I, DIVISIONS 1 AND 2.
Not a product of Appleton.

NEC® 501-14. SIGNALING, ALARM, REMOTE-CONTROL AND COMMUNICATION SYSTEMS.
17—INTRAGROUND™ N1 SERIES PILOT LIGHT (Section M); sealing Unilets® required.

NEC® 501-15. LIVE PARTS, CLASS I, DIVISIONS 1 AND 2.
There shall be no exposed live parts.

NEC® 501-16. GROUNDING, CLASS I, DIVISIONS 1 AND 2.
The use of Appleton SD Static Discharge Reels (Section Z) is recommended for use with metal parts of equipment such as frames, motors, machinery, and other generators of electricity. See pages 83 and 94 of this booklet.

NOTE: Diagram is based on the assumption that 2″ or larger conduit is not used, except for panelboards. Item 9 in the diagram is a non-factory sealed panelboard with main conduit hubs larger than 2″ and therefore requires seals per NEC® 501-5(a)(1)(2). (Item 9A is a factory-sealed panelboard, but external seals are required because conduit to the main conduit hub is larger than 2″.)

Figure 4–3 *(continued)* Power Diagram for a Class I Division 1 location.

- Locations where flammable liquids or gases may be present in ignitable concentrations during normal operations
  (See Figures 4–2, 4–3 and 4–4)

CLASS I DIVISION 2

Class I Division 2 locations are those in which volatile flammable liquids or gases are handled, processed, or used, but in which they will normally be confined within closed containers or closed systems from which they can escape only in the case of accidental rupture or breakdown of the containers or systems (see Figure 4–5). The hazardous conditions will occur only under abnormal conditions. Class I areas may also be classified Division 2 where ignitable concentrations of flammable gases or vapors are prevented by positive mechanical ventilation where safeguards against ventilated failure are provided.

Class I Hazardous (Classified) Areas 41

Figure 4–4 Lighting Diagram for a Class I Division 1 location. (Courtesy of Appleton Electric Company)

CLASS I, GROUPS A or B, DIVISION 1 INSTALLATIONS

NEC® 501-1. GENERAL
The general rules of this code shall apply to the installation of electrical wiring and equipment in locations classified as Class I UNDER Section 500-5 except as modified by this Article.

NEC® 501-2. TRANSFORMERS AND CAPACITORS.
Not a product of Appleton.

NEC® 501-3. METERS, INSTRUMENTS AND RELAYS.

NEC® 501-4. WIRING METHODS.
These products are suitable for Groups A and B.
1—GRH and GRF CAST CONDUIT OUTLET BOX UNILETS (Section J).
1A—GRSS CAST CONDUIT OUTLET BOX UNILETS® (Section J).
2—UNF UNIONS (Section I).
3—ELMFL ELBOWS (Section I).
4—EXGJH FLEXIBLE COUPLINGS (Section I).

NEC® 501-5. SEALING.
5—EYS VERTICAL SEALING UNILETS: suitable for Groups A and B (Section I).

NEC® 501-6. SWITCHES, CIRCUIT BREAKERS, MOTOR CONTROLLERS AND FUSES.
6—EDS PUSH BUTTON CONTROL STATION, FACTORY SEALED FOR GROUP B; sealing Unilets® not required.
6A—XSE (OR SOME ES OR ETS) MAGNETIC MOTOR STARTER; sealing Unilets® required.
7—EDS SWITCH, FACTORY SEALED FOR GROUP B; sealing Unilets® not required (Section N).
8—OFC CONTROL STATION, SUITABLE FOR GROUP B; sealing Unilets® required (Section M).

NEC® 501-7. CONTROL TRANSFORMERS AND RESISTORS.
Not a Standard Catalog listing.

NEC® 501-8. MOTORS AND GENERATORS.
Not a standard product listing but motors provided with some Reelite® products (Section W3).

NEC® 501-9. LIGHTING FIXTURES.
9—A-51 INCANDESCENT OR HID, FACTORY SEALED FOR GROUPS A AND B; sealing Unilets® not required (Section G1 & G2).
9A—CODE•MASTER JR. HID FACTORY SEALED LIGHTING FIXTURE; sealing Unilet® required for Group B (Section G1).
10—EFHCA FIXTURE HANGERS (Section H).

NEC® 501-10. UTILIZATION EQUIPMENT, FIXED AND PORTABLE.
UTILIZATION EQUIPMENT; Products such as portable lamps must be approved for Class I, Groups A and B, Division 1 locations.
11—ECC FACTORY SEALED CORD CONNECTORS (Section L).
The Appleton ECC is an approved method for permanent connection from a rigid conduit system to a portable appliance. Supplies power to utilization equipment.

NEC® 501-11. FLEXIBLE CORDS.
Not a standard product listing, but Flexible cord is provided with Appleton Reelites designed for hazardous locations (Section W3).

NEC® 501-12. RECEPTACLES AND ATTACHMENT PLUGS.
12—U-LINE® PLUGS AND RECEPTACLES, FACTORY SEALED FOR GROUP B; sealing Unilets® not required (Section L).
13—FSQX and EBRH PLUGS AND RECEPTACLES, SUITABLE FOR GROUP B; sealing Unilets® required (Section L).

NEC® 501-13. CONDUCTOR INSULATION, CLASS I, DIVISIONS 1 AND 2.
Not a product of Appleton.

NEC® 501-14. SIGNALING, ALARM, REMOTE-CONTROL AND COMMUNICATION SYSTEMS.

NEC® 501-15. LIVE PARTS, CLASS I, DIVISIONS 1 AND 2.
There shall be no exposed live parts.

NEC® 501-16. GROUNDING, CLASS I, DIVISIONS 1 AND 2.
The use of Appleton SD Static Discharge Reels (Section Z) is recommended for use with metal parts of equipment such as frames, motors, machinery, and other generators of electricity. See pages 83 and 94 of this booklet.

Figure 4–4 *(continued)* Lighting Diagram for a Class I Division 1 location.

Figure 4–5 Example of Class I Division 2 location lighting in a refinery. (Courtesy of Crouse-Hinds, Division of Cooper Industries)

"Electrical conduits and their associated enclosures separated from process fluids by a single seal or barrier shall be classed as a Division 2 location if the outside of the conduit and enclosures is an unclassified location."

Examples of Class I Division 2 Locations

- Locations where volatile flammable liquids, gases, or vapors are used but in the judgment of the authority having jurisdiction would become hazardous in the case of an accident or unusual operating condition (see Figures 4–6 and 4–7)
- Locations where volatile flammable liquids, gases, or vapors are used but in the judgment of the authority having jurisdiction adequate ventilating equipment with safeguards are

Class I Hazardous (Classified) Areas 43

Figure 4–6 Lighting Diagram for a Class I Division 2 location. (Courtesy of Appleton Electric Company)

CLASS I, DIVISION 2

NEC® 501-1. GENERAL
The general rules of this Code shall apply to the installation of electrical wiring and equipment in locations classified as Class I under Section 500-5 except as modified by this Article.

NEC® 501-2. TRANSFORMERS AND CAPACITORS.
Not a product of Appleton.

NEC® 501-3. METERS, INSTRUMENTS AND RELAYS.

NEC® 501-4. WIRING METHODS.
1—LL and T TYPE CAST CONDUIT OUTLET BODIES (Section A).
1A—JB ROUND BASE CAST CONDUIT OUTLET BOX UNILETS® (Section A).
1B—GS CAST CONDUIT OUTLET BOX UNILETS® (Section A).
1C—FS OR FD WITH BLANK CAST COVER (Section B).
2—LBD/LBDN PULLING UNILET® (Section A).
3—UNF(NR) UNIONS AND ELF ELBOWS (Section I).
4—FLEXIBLE COUPLINGS (Section I).

NEC® 501-5. SEALING.
5—EYS VERTICAL SEALING UNILETS® (Section I).
5A—ESU OR EY HORIZONTAL SEALING UNILETS® (Section I).
6—ECDB DRAIN/BREATHERS (Section I).

NEC® 501-6. SWITCHES, CIRCUIT BREAKERS, MOTOR CONTROLLERS AND FUSES.
7—D2P FACTORY SEALED PANELBOARDS (Section P). Also refer to *NEC®* Art. 430.
8—EB CIRCUIT BREAKER UNILETS® (Section O). Also refer to *NEC®* Art. 430.
9—EDS FACTORY SEALED SWITCH UNILETS® (Section N); requires no sealing Unilets®
10—GUSC SWITCH (Section N). Also refer to *NEC®* Art. 430.
10A—EFD FACTORY SEALED PUSH BUTTON CONTROL STATION (Section M); requires no sealing Unilet® Also refer to *NEC®* Art. 430.

NEC® 501-7. CONTROL TRANSFORMERS AND RESISTORS.
Not a standard catalog listing—Unilet® modifications can be made.

NEC® 501-8. MOTORS AND GENERATORS.
Not a standard product listing, but motors provided with some Reelite® products.

NEC® 501-9. LIGHTING FIXTURES.
12—STYLMASTER ENCLOSED AND GASKETED INCANDESCENT LIGHTING FIXTURE (Section F2).
12A—MERCMASTER ENCLOSED AND GASKETED HID LIGHTING FIXTURE (Section F1).
12B—MERCMASTER III ENCLOSED AND GASKETED HID LIGHTING FIXTURE (Section F1).
12C—VRS ENCLOSED AND GASKETED FLUORESCENT LIGHTING FIXTURE (Section G3).
12D—EHL FACTORY SEALED EXPLOSION-PROOF HANDLAMP (Section G2).
12E—MERCMASTER III LOW PROFILE ENCLOSED AND GASKETED HID LIGHTING FIXTURE (Section F1).
12F—AREAMASTER 250/400 FLOODLIGHT with Yoke Mount (Section Q).
13—AHG FLEXIBLE FIXTURE HANGER (Section E).
13A—JB AND GS FIXTURE HANGER (Section E).
13B—FHS DUMMY CONDUIT SUPPORT (Sections E & H).
13C—WALL MOUNT for Mercmaster III (Section F1).
13D—LIQUIDTIGHT ST CONNECTOR For flexible metal conduit (Section I).

NEC® 501-10. UTILIZATION EQUIPMENT, FIXED AND PORTABLE.
UTILIZATION EQUIPMENT—Products such as portable lamps must be approved for Class I locations.
14—ECC FACTORY SEALED CORD CONNECTORS (Section L).
The Appleton ECC is an approved method for a permanent connection from a rigid conduit system to a portable appliance. Supplies power to utilization equipment.
15—RL200E FACTORY SEALED REELITES® (Section W3). Available with EHL explosion-proof handlamp.

NEC® 501-11. FLEXIBLE CORDS, CLASS I, DIVISIONS 1 AND 2.
Not a standard product listing, but flexible cord is provided with Appleton Reelites® (Section W3) designed for hazardous locations.

NEC® 501-12. RECEPTACLES AND ATTACHMENT PLUGS, CLASS I, DIVISIONS 1 AND 2.
16—EFS U-LINE® FACTORY SEALED PLUGS AND RECEPTACLES (Section L).
16A—FSQX RECEPTACLE with interlocking safety switch (Section L).

NEC® 501-13. CONDUCTOR INSULATION, CLASS I, DIVISIONS 1 AND 2.
Not a product of Appleton.

NEC® 501-14. SIGNALING, ALARM, REMOTE-CONTROL AND COMMUNICATION SYSTEMS.
17—INTRAGROUND™ N2 SERIES FACTORY SEALED PILOT LIGHTS (Section M); sealing Unilets® not required.

NEC® 501-15. LIVE PARTS, CLASS I, DIVISIONS 1 AND 2.
There shall be no exposed live parts.

NEC® 501-16. GROUNDING, CLASS I, DIVISIONS 1 AND 2.
The use of Appleton SD Static Discharge Reels (Section Z) is recommended for use with metal parts of equipment such as frames, motors, machinery, and other generators of electricity. See pages 83 and 94 of this booklet.

Figure 4–6 *(continued)* Lighting Diagram for a Class I Division 2 location.

Class I Hazardous (Classified) Areas 45

Figure 4–7 Power Diagram for a Class I Division 2 location. (Courtesy of Appleton Electric Company)

CLASS I, DIVISION 2

NEC® 501-1. GENERAL
The general rules of this Code shall apply to the installation of electrical wiring and equipment in locations classified as Class I under Section 500-5 except as modified by this Article.

NEC® 501-2. TRANSFORMERS AND CAPACITORS.
Not a product of Appleton.

NEC® 501-3. METERS, INSTRUMENTS AND RELAYS.
1—GUB/GUBB EXPLOSION-PROOF ENCLOSURES (Section K).
1A—GUBM/GUBBM EXPLOSION-PROOF ENCLOSURE (Section K) with glass for viewing of instruments.

NEC® 501-4. WIRING METHODS.
2—T and LR TYPE CAST CONDUIT OUTLET BODIES (Section A).
2A—NEC® 6X8X SERIES MOGUL CAST CONDUIT OUTLET BODIES (Section A).
2C—JB or GS CAST CONDUIT OUTLET BOXES (Section A).
3—RS CAST JUNCTION BOX (Section C).
4—UNF and UNF(NR) UNIONS (Section I).
5—ELF ELBOWS (Section I).
6—EXGJF FLEXIBLE COUPLINGS (Section I).

NEC® 501-5. SEALING.
7—ESU SEALING UNILETS® (Section I).
8—ECDB DRAIN/BREATHERS (Section I).

NEC® 501-6. SWITCHES, CIRCUIT BREAKERS, MOTOR CONTROLLERS AND FUSES.
9—D2P FACTORY SEALED PANELBOARDS (Section P). Also refer to NEC® Art. 430.
9A—EB CIRCUIT BREAKER UNILETS® (Section O). Also refer to NEC® Art. 430.
10—ETC COMBINATION CIRCUIT BREAKERS AND MOTOR STARTERS (Section O). Also refer to NEC® Art. 430.
11—INTRAGROUND™ N2 SERIES FACTORY SEALED PUSH BUTTON CONTROL STATION UNILETS® (Section M); sealing Unilets® not required. Also refer to NEC® Art. 430.
12—GUSC SWITCH UNILET® (Section N). Also refer to NEC® Art. 430.
13-EFS SWITCH UNILETS® (Section N).

NEC® 501-7. CONTROL TRANSFORMERS AND RESISTORS.
14—GUB/GUBB EXPLOSION-PROOF UNILET® ENCLOSURE (Section K).

NEC® 501-8. MOTORS AND GENERATORS.
Not a standard product listing, but motors provided with some Reelite® (Section W3) products.

NEC® 501-9. LIGHTING FIXTURES.
Best described on next page.

NEC® 501-10. UTILIZATION EQUIPMENT, FIXED AND PORTABLE.
UTILIZATION EQUIPMENT—Products such as portable lamps must be approved for Class I locations.
15—ECC FACTORY SEALED CORD CONNECTORS (Section L). Appleton ECC is an approved method for a permanent connection from a rigid conduit system to a portable appliance. Supplies power to utilization equipment.

NEC® 501-11. FLEXIBLE CORDS, CLASS I, DIVISIONS 1 AND 2.
Not a standard product listing, but flexible cord is provided with Appleton Reelites® (Section W3) designed for hazardous locations.

NEC® 501-12. RECEPTACLES AND ATTACHMENT PLUGS, CLASS I, DIVISIONS 1 AND 2.
16A—CES FACTORY SEALED PLUGS AND RECEPTACLES (Section L); sealing Unilets® not required.
16B—FSQX PLUGS AND RECEPTACLES (Section L); sealing Unilets® required.
16C—EBR PLUGS AND RECEPTACLES (Section L) with manual disconnect switch; sealing Unilets® not required.

NEC® 501-13. CONDUCTOR INSULATION, CLASS I, DIVISIONS 1 AND 2.
Not a product of Appleton.

NEC® 501-14. SIGNALING, ALARM, REMOTE-CONTROL AND COMMUNICATION SYSTEMS.
17—INTRAGROUND™ N2 SERIES FACTORY SEALED PILOT LIGHT (Section M); sealing Unilets® not required.

NEC® 501-15. LIVE PARTS, CLASS I, DIVISIONS 1 AND 2.
There shall be no exposed live parts.

NEC® 501-16. GROUNDING, CLASS I, DIVISIONS 1 AND 2.
The use of Appleton SD Static Discharge Reels (Section Z) is recommended for use with metal parts of equipment such as frames, motors, machinery, and other generators of electricity. See pages 83 and 94 of this booklet.

Figure 4–7 *(continued)* Power Diagram for a Class I Division 2 location.

used to ensure that the ventilation equipment will provide continuous ventilation air movement

Factors for Determining Division 2 Areas

- Limited quantity of flammable material that might escape in an accident or spill
- Adequacy of ventilating equipment
- Specific industry or business records, as they relate to previous explosions and fires

Equipment for Class I Division 1 Locations

Equipment used in Class I Division 1 locations must be marked in accordance with *NEC®* Section 500–3(d) with the following:

- Class
- Group
- Operating temperature or temperature range referenced to 40 degrees Celsius Ambient.

Devices for Class I locations are housed in enclosures which are designed strong enough to contain an explosion if the hazardous vapors enter the enclosure and are ignited. These enclosures then cool and vent the products of combustion in such a way that the surrounding atmosphere is not ignited. It is important to note that equipment designed and listed or approved to be installed in these locations is *not* "gas tight." It *is* designed to contain the explosive forces and internal pressures if an explosion should occur.

Heat-producing equipment for hazardous locations, such as lighting fixtures, must not only contain the explosion and vent the cooled products of combustion, but must also be designed to comply with *NEC®* Section 500–3(e) and operate with surface temperatures below the ignition temperatures of the hazardous atmosphere (see Figures 4–4 and 4–8).

Specific Equipment for Division 1 Locations

Transformers and capacitors which do not contain liquids that will burn must be installed in vaults or be approved for Class I installations. Transformers and capacitors which do contain liquids that burn must be installed in vaults complying with *NEC®* 450–41 through 450–48 with adequate ventilation for the continuous removal of flammable gases and vapors. Sufficient venting is required to relieve explosion pressures if an explosion should occur. Doors or openings are not permitted between the Class I area and the vault (see Figure 4–3).

Meters, instruments, and relays, such as kilowatt-hour meters, thermionic tubes, instrument transformers, must be provided with enclosures approved for Class I Division 1 locations.

Switches, motor controllers, circuit breakers, fuses, push buttons, relays, and similar devices must be provided with enclosures and together they must be approved as a complete assembly for the Class I Division 1 location. Control transformers, impedance coils, and resistors including associated switching mechanisms and their enclosures must be approved as a complete assembly for the Class I Division 1 location (see Figure 4–3).

Motors, generators, and other rotating electric machinery must be approved for Class I Division 1 locations or meet the special manufacturing and installation requirements as specified in *NEC®* Section 501–8(a).

Lighting fixtures must be approved as a complete assembly for Class I Division 1 locations and clearly marked for the maximum lamp wattage permitted. All fixtures are required to have physical protection either by location or suitable guards. All boxes, box assemblies, or fittings used to support fixtures must be approved for Class I Division 1 locations. Portable fixtures must be approved for portable use and comply with the above requirements (see Figures 4–9 and 4–10).

Design of Explosion-Proof Equipment

Threaded Joint Construction

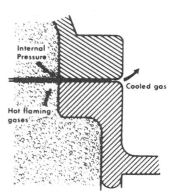

Figure 4–8 Explosion-proof construction is not gas-tight by design. It is designed and manufactured to contain an explosion and to prevent the escape of fire and hot gases. (Courtesy of Appleton Electric Company)

Figure 4–9 Explosion-proof light fixtures. (Courtesy of Crouse-Hinds, Division of Cooper Industries)

EQUIPMENT FOR CLASS I DIVISION 2 LOCATIONS

Equipment used in Class I Division 2 locations must be marked in accordance with *NEC®* Section 500–3(d) with the following:

- Class
- Group
- Operating temperature or temperature range referenced to 40 degrees Celsius Ambient

NEC® 500–3(d) Exception 2. Fixed lighting fixtures are not required to be marked to indicate the group.
NEC® 500–3(d) Exception 3. Fixed general purpose equipment except fixed lighting fixtures, acceptable for Class I Division 2 are not required to be marked with Class, Group, Division, or operating temperature.

Figure 4–10 Portable reel-lights suitable for use in Class I Divisions 1 & 2. (Courtesy of Appleton Electric Company)

Devices for Class I Division 2 locations are housed in enclosures which are designed strong enough to contain an explosion if the hazardous vapors enter the enclosure and are ignited. These enclosures then cool and vent the products of combustion in such a way that the surrounding atmosphere is not ignited. It is important to note that equipment designed and listed or approved to be installed in these locations is *not* "gas tight." It *is* designed to contain the explosive forces and internal pressures if an explosion should occur (see Figure 4–7). However, if all of the provisions of Section 501–6(b)(1) through (4) are met, using general purpose enclosures to house certain equipment may be permitted.

Specific Equipment for Division 2 Locations

Transformers and capacitors must comply with *NEC®* Sections 450–21 through 450–27. Meters, instruments, and relays, such as kilowatt-hour meters, thermionic tubes, instrument transformers, must be provided with enclosures approved for Class I Division 1 locations.

Switches, circuit breakers, fuses, make and break contacts of push buttons, relays, and similar devices located in a Class I Division 2 area must meet the requirements of Section 501–3(b)(2) and must be provided with enclosures and together they must be approved as a complete assembly for the Class I Division 1 location. Control transformers, impedance coils, and resistors together with switching mechanisms and their enclosures must be approved as a complete assembly for the Class I Division 1 location. General purpose enclosures are permitted provided the current-interrupting contacts are immersed in oil, enclosed within a chamber which is hermetically sealed to prevent the entrance of gases or vapors or in nonincendive circuits which under normal conditions do not release enough energy to ignite the specific ignitable atmospheric mixture (see Figure 4–7).

Motor controllers, motors, generators, and other rotating electric machinery located in Class I Division 2 areas must be approved for Class I Division 1 locations or meet the special manufacturing and installation requirements as specified in *NEC®* Section 501–8(b).

Equipment used in Class I Division 2 locations must be marked in accordance with *NEC®* Section 500–3(d) with the following:

- Class
- Group
- Operating temperature or temperature range referenced to 40° Celsius Ambient

WIRING METHODS

The minimum wiring methods are specified in *NEC®* Article 501 Section 501–4(a) for Class I Division 1 locations. *NEC®* Section 501–4(b) covers Class I Division 2 locations.

Class I Division 1

NEC® 501–4(a) requires wiring methods of *NEC®* Article 346 Threaded Rigid Metal Conduit,

Figure 4–11 Explosion-proof flexible conduit is approved for limited use where necessary for flexibility. Use only approved wiring methods pictured above. (Courtesy of Crouse-Hinds, Division of Cooper Industries)

NEC® Article 345 Threaded Steel Intermediate Metal Conduit (IMC), or *NEC®* Article 330 type MI cable with termination fittings approved for the location. All boxes, fittings, and joints shall be threaded for a connection to conduit or cable terminations and shall be explosion-proof. Threaded joints must be made up with at least five threads fully engaged (see Figures 4–9, 4–11, and 4–13).

> MI cable must be installed and supported in a manner to avoid tensile stress at the terminations.

Where necessary to use flexible connections as provided in *NEC®* Section 501–11, such as for motor terminals, flexible fittings approved for Class I locations must be used (see Figure 4–10). *NEC®* Section 501–11, however, permits flexible cord or connection between portable lighting equipment or other portable utilization equipment and a fixed portion of its supply circuit with rigid requirements. Extra-hard usage cord must be employed and must, in addition to the conductors in the circuit, employ a grounding conductor complying with *NEC®* Section 400–23. It must be connected to terminals or supply conductors in an approved manner and be supported by clamps or other suitable means in such manner that there will be no tension on the terminal connections (see Figure 4–12). The cord and connections must be provided with seals where the cord enters boxes, fittings, or enclosures of explosion-proof type (see Figure 4–12).

In industrial establishments with restricted access and where conditions of maintenance and supervision assure that only qualified persons will service the installation, type MC cable listed for the use (Class I Division 1 locations) with terminations is permitted.

Figure 4–12 Electrical equipment with a static ground clamp. (Courtesy of Crouse-Hinds, Division of Cooper Industries)

> This product will be developed now that it has been accepted into the 1996 *NEC*®. It will be a special type of MC cable which employs a gas/vaportight continuous aluminum sheath and an overall jacket of suitable polymeric material and a separate grounding conductor complying with Section 250–95.

Figure 4–13 The example of a thread lubricant manufactured for the purpose; only approved thread lubricants should be used. (Courtesy of Crouse-Hinds, Division of Cooper Industries)

Rigid nonmetallic conduit (Article 347) is permitted where encased in a minimum of 2 inches of concrete and buried below the surface at least 2 feet. Threaded rigid metal conduit or intermediate metal conduit shall be used for the last 2 feet of the run and to the point of emergence or point of connection to aboveground raceways. An equipment-grounding conductor shall be provided in accordance with Section 250–95 to provide continuity of the raceway system.

Class I Division 2

NEC® Section 501–4(b) requires somewhat less stringent wiring requirements. It is equally as important, however, to wire areas classified as Class I Division 2 correctly. The Class I Division 2 minimum wiring methods are as follows: *NEC*® Article 346 Threaded Rigid Metal Conduit, *NEC*® Article 345 Threaded Steel Intermediate Metal Conduit, *NEC*® Article 364 Enclosed Gasketed Busways, *NEC*® Article 362 Enclosed Gasketed Wireways, PLTC cable in accordance with the provisions of *NEC*® Article 725 and types MI, MC, MV, PC, or SNM cable with approved fittings and terminations are permitted.

> Types PLTC, MI, MC, MV, TC, and SNM are permitted to be installed in cable trays and shall be installed in a manner to avoid tensile stress at the termination fittings.

Boxes and joints are not required to be explosion-proof in these areas, except as required in *NEC*® Sections 300(b)(1), 501–6(b)(1), and 501–14(b)(1). When provisions must be made for limited flexibility, such as at motor terminals, flexible metal conduit with approved fittings, *NEC*® Article 333 Type AC Armored Cable with approved fittings, *NEC*® Article 351 Liquidtight Flexible Metal Conduit with approved fittings, and Liquidtight Flexible Nonmetallic Conduit with approved fittings, or flexible cord approved for hard uses provided with approved bushed fittings may be used (see Figure 4–12). The additional conductor for grounding must be included in the flexible cord unless other accessible means for grounding are provided.

Wiring which cannot release sufficient energy to ignite a specific ignitable atmospheric mixture by opening, shorting, or grounding may be permitted in this area, using any of the methods suitable for general wiring applications.

Class I Hazardous (Classified) Areas

> **Nonincendive circuits** are permitted in wiring methods suitable for ordinary locations.

SEALING AND DRAINAGE REQUIREMENTS

NEC® Section 505 covers the minimum requirements for sealing and drainage in Class I Divisions 1 and 2. Fittings approved for the location and purpose must be used with approved sealing compounds; sealing compounds approved for type MI cable must also exclude moisture and other fluids from the cable insulation. In addition, see the requirements of *NEC®* Section 505–5(f)(3).

> Seals are provided in conduit and cable systems to minimize the passage of gases and vapors and prevent the passage of flames through the wiring method from one electrical system or enclosure to another. Type MI cable inherently prevents the passage of gases and vapors through the insulation, therefore seals are not required.

The specific requirements for installing the seals are found in *NEC®* Section 501–5(c) and in the manufacturer's instructions. Installation of these seals is a critical part of this installation and improperly made will not provide the protection needed to isolate accidental explosions. Raceways in Class I hazardous locations are required to be provided with sealing fittings to prevent the passage of gases, vapors, or flames through the raceway from one portion of an electrical installation to another and to prevent pressure piling or cascading. Pressure piling may occur in these locations when an explosion in some part of the system, such as a switch enclosure, is not effectively blocked or sealed off. The first explosion may be made harmless if properly confined by a seal to the switch enclosure. If allowed to enter the raceway, it builds up great pressure, and another more powerful explosion may follow. The second explosion builds up additional pressures in the gas-filled conduit and a greater explosion results. This may continue in very long raceway runs not provided with intermittent seals until some part of the system ruptures. The hot or burning gases could then escape and ignite flammable gases in the surrounding atmosphere, causing a general explosion.

Proper sealing is one of the most important requirements for safety in a Class I hazardous location. Explosive vapors are prevented from passing to a nonhazardous area where they could be ignited and carry the flame back to the explosive vapor location. In the manner of a pilot light on a gas range, arcs, sparks, and hot particles from an arcing device are also prevented from entering the raceway by the seal. Sealing fittings are designed to permit the pouring of an approved water-base sealing compound into the fitting around and between the conductors (see Figure 4–16).

> Conduit seals are limited to a 25% conductor fill by their listing. Where raceways are filled to 40%, check with the manufacturer for conductor fill limits (see Figures 4–14 and 4–15).

The raceway should first be blocked at the bottom of the sealing fitting with a packed material to prevent the sealing mixture from draining out of the fitting. The liquid mixture solidifies after a few hours and becomes very hard. Proper mixing of the compound and liquid is essential; otherwise the seal would not prevent the pressure piling. Even though the seal fitting may prevent the passage of vapors, arcs and hot particles, all voids under, between, and over conductors must be filled if the sealing mixture is to perform its function of preventing the passage of gases, vapors, or flames through the raceway. The elimination of the pressure piling can be achieved by installing effective seals in every raceway regardless of size where it enters or leaves any junction box and at regular intervals. Seals should also be installed in all raceway runs. The minimum installation is to install a seal at all points in the system where they are required by the *NEC®*. In the absence of tests for various gases, to determine the proper spacing of seals, in long conduit runs it is not unreasonable to specify that they be installed at 50-foot intervals in all conduit runs within a Class I Division 1 location.

If installed in accordance with the *NEC®* requirements, a rigid steel conduit system with suitable enclosure and seals will effectively contain an internal explosion and prevent the escape of burning gases until they are cooled sufficiently to prevent the ignition of flammable vapors in the surrounding atmosphere. The gases are cooled by expansion as they escape through the machine-grounded joints of predetermined width

Figure 4–14 Examples of fittings that are approved for use in hazardous locations. (Courtesy of Crouse-Hinds, Division of Cooper Industries)

and through threaded joints. Ground joints, unless properly maintained, are not as reliable as threaded joints. A single loose screw in a cover of a ground joint, or scratches in ground services, will render an otherwise explosion-proof enclosure ineffective. Threaded joints, on the other hand, are reliable because the gases escape only by following the threads. The *NEC*® requires at least five full threads be engaged at each coupling. Effective seals must be placed in a raceway as required by the *NEC*® within 18 inches of

Figure 4–15 Cutaway view of a new seal fitting which is sized for a 40% fill the same as all typical raceways approved for installation in Class I locations. (Courtesy of Appleton Electric Company)

an arcing device or at or near the point where the raceway leaves or enters a hazardous area. In the case of a raceway two inches in diameter or larger, seals are required within 18 inches of any junction box containing splices, taps, or terminal connections. No coupling or fitting is permitted between the seal and the boundary between the hazardous and nonhazardous areas. In many cases, special panelboards and controls are designed with factory seals that are an integral part of the equipment designed for Class I locations. Because such seals effectively isolate the arc-producing segments of the assembly, additional seals are not generally required in the conduit leaving these panelboards. Incandescent and fluorescent fixtures in Class I locations are also available with integral seals (see Figure 4–17).

Figure 4–16 Examples of sealing fittings. Methods for damming and pouring sealing compounds. (Courtesy of Appleton Electric Company)

Figure 4–17 Examples of factory sealed (A) pilot light and (B, C) push-button stations. (A and B courtesy of Appleton Electric Company; C courtesy of Crouse-Hinds, Division of Cooper Industries)

> Seals are not intended to prevent the passage of gases or vapors at a continuous pressure. Where there are differences in pressure across the seal, there may be some passage of gases or vapors.

> Leakage and the passage or propagation of flames through the interstices of stranded conductors can be reduced through the use of compact stranding or sealing of the individual stranding.

Although wiring methods are less restrictive in a Class I Division 2 location, sealing is still required as specified in *NEC®* Section 501–5(b). In each conduit run passing from a Class I Division 2 location into an unclassified area, a seal fitting is required.

- This sealing fitting may be located on either side of the boundary. It must be designed and installed to minimize the amount of gas or vapor that may enter the conduit system within the Class II location.
- It must also prevent the gas or vapor from being communicated to the conduit beyond the seal.
- Rigid metal conduit or threaded steel intermediate conduit must be used between the sealing fitting and the point at which the conduit leaves the Division 2 location.
- A threaded connection shall be used at the sealing fitting.
- Fittings or couplings are not permitted between the sealing fitting and the point in which the conduit leaves the Division 2 location.
- Conduit systems terminating to an open raceway in an outdoor unclassified area are not required to be sealed between the point at which the conduit leaves the classified location and enters the open raceway.
- The minimum requirements for the location and installation of sealing fittings are found in *NEC®* Section 501–5(a), (b), (c), (d), (e), and (f).

> Special type seals are available for retrofitting seals on existing installations. Cable fittings are available for the new type MC explosion-proof cable which will meet the new Exception 2 to Section 501–4(a) (see Figures 4–18 and 4–19).

Drainage

Where it is likely that water or other condensed vapors may be trapped in the system equipment enclosures or in the raceways, an approved means

Class I Hazardous (Classified) Areas 55

Figure 4–18 Examples of fittings for the new type MC cable that will be permitted in accordance with Section 501–4(a) Exception 2. (Courtesy of Appleton Electric Company)

Figure 4–19 Example of a new type split seal fitting that can be used in existing Division 2 locations when needed.

must be provided to prevent such accumulation or provide a means for the drainage of those fluids or vapors periodically (see Figure 4–20). An integral means for periodic drainage shall be provided by the manufacturer of motors and generators where needed or the conduit system and equipment shall be arranged to minimize the entrance of liquids and vapors acceptable to the authority having jurisdiction. Equipment, such as canned pumps, process or service connections for flow, pressure, analysis measurement, etc., that depend upon a tube, single compression seal, or diaphragm to prevent flammable or combustible fluids from entering the raceway system, must also have an additional approved seal, barrier, or other means to prevent propagation if the primary seal should fail. The additional redundant means must be approved for the temperature and pressures which they may be subjected to if a primary means should fail.

Drains, vents, or other devices are required so that the primary seal leakage will be obvious.

Figure 4–20 Example of seals and drain and vent fittings. (Courtesy of Appleton Electric Company)

9. Extra-hard usage cord is only approved in Class I Division 1 locations.
   Answer: True or False.   Reference:

10. Conduit runs which pass through hazardous Class I locations and do not have enclosures within 12 inches of the boundary are not required to have seals installed.
    Answer: True or False.   Reference:

Chapter 5

Class II

CLASS II LOCATIONS

Class II locations are those that are hazardous because of the presence of combustible dust.

Class II Location Examples

- Grain elevators and bulk handling facilities
- Magnesium manufacture and storage
- Starch manufacture and storage
- Fireworks manufacture and storage
- Flour and feed mills
- Areas for packaging and handling of pulverized sugar and cocoa
- Facilities for the manufacture of magnesium and aluminum powder
- Some coal preparation plants and coal handling facilities
- Spice grinding plants
- Confectionery manufacturing plants
  (See Figures 5–1 and 5–2.)

> See NFPA 497B and NFPA 497M for other hazardous (classified), locations, such as agriculture and chemical process areas.

Class II Group Classifications

Dusts have also been placed in groups designated E, F, and G, based on their particular hazardous characteristics and the dusts' electrical resistivity. It is important to select equipment suitable for the specific hazardous group to insure that the installation is as safe as possible.

Figure 5–1 Class II grain elevator locations. (Courtesy of Appleton Electric Company)

Figure 5–2 Grain storage silos can be extremely dangerous. It may be possible to avoid disasters like this by eliminating common sources of ignition. (Courtesy of Appleton Electric Company)

> **Note:** When the division is not specified, the equipment is suitable for both Division 1 and 2 applications.

> **Warning:** Where Class II Group E dusts are present in hazardous quantities this area or areas are classified Class I Division 1. There are no Division 2 areas.
> Dusts containing magnesium or aluminum dust are particularly hazardous; *extreme caution* is advised to avoid ignition and explosion.

CLASS II DIVISION 1

Class II locations are areas where the combustible dust may be in suspension in the air under normal conditions and in sufficient quantities to produce explosive or ignitable mixtures. The classification of the area would apply where the combustible dusts are likely to be present continuously, intermittently, or periodically. Division 1 locations also exist where failure or malfunction of machinery or equipment might cause a hazardous location to exist while providing a source of ignition with the simultaneous failure of electrical equipment. Included also are locations in which combustible dust of an electrically conductive nature may be present.

Class II Division 1 Location Examples

- Areas used for the processing and handling of grain products
- Areas used for the processing and handling of cocoa, sugar, egg, and milk powders
- Areas used for the processing and handling of pulverized spices, starch, and paste
- Areas used for the processing and handling of potato and wood flour
- Areas used for the processing and handling of oil meal from seeds or beans
- Areas used for the processing and handling of dusts containing aluminum or magnesium

CLASS II DIVISION 2

A Class II Division 2 location is one in which combustible dust will not normally be in suspension in the air, and normal operations will not put the dust in suspension in the air, but where accumulation of the dust may interfere with the safe dissipation of heat from electrical equipment or where accumulations near electrical equipment may be ignited by arcs, sparks, or burning material from the equipment.

Class II Division 2 Examples

- Grain elevators and bulk handling facilities
- Magnesium storage
- Starch storage
- Fireworks storage
- Areas within flour and feed mills
- Areas for storing pulverized sugar and cocoa
- Aluminum powder storage areas
- Areas in some coal preparation plants and coal handling facilities
- Areas for storing spices in grinding plants
- Storage areas in confectionery manufacturing plants

Figure 5–3 Class II lighting installation. (Courtesy of Crouse-Hinds, Division of Cooper Industries)

Note: The authority having jurisdiction may determine that areas can be classified Division 2 or considered as nonhazardous areas if adequate dust removal systems are installed and maintained or where products such as seed are handled in a manner which assures limitation of dust quantities to warrant declassification to a lower class or in some instances to an unclassified area. For guidance, see NFPA 497B and 497M.

EQUIPMENT FOR CLASS II LOCATIONS

The enclosures used to house devices in Class II locations are designed to seal out dust. Contact between the hazardous atmosphere and the source of ignition has been eliminated and no explosion can occur within the enclosure (see Figures 5–3 and 5–4).

Equipment and wiring of the type defined in Article 100 as explosion-proof shall not be required and shall not be acceptable in Class II locations unless approved for such locations.

As in Class I equipment, heat-producing equipment must be designed to operate below the ignition temperature of the hazardous atmosphere. However, in Class II equipment, additional consideration must be given to the heat buildup that may result from the layer of dust which will settle on the equipment.

SPECIFIC EQUIPMENT FOR CLASS II LOCATIONS

Class II Division 1 Locations

Switches, circuit breakers, motor controllers, and fuses including push buttons, relays, and similar devices that interrupt current during normal operation, or are installed where electrically conductive combustible dusts may be present, must be installed in dust-ignitionproof enclosures. The enclosed equipment must be listed or approved as a complete assembly for Class II locations.

In areas where electrical combustible dusts are not present, switches containing no fuses and not designed or intended to interrupt current may be enclosed within metal enclosures that shall be designed to minimize the entrance of dust and that shall be equipped with telescoping or close-fitting covers or with other acceptable means to prevent the escape of sparks, burning material, or to allow the entrance of combustible dusts (see Figure 5–5). They shall have no openings (i.e., holes for attachment screws).

Figure 5–4 Example of a Class II switched receptacle. (Courtesy of Crouse-Hinds, Division of Cooper Industries)

Figure 5–5 Example of a conduit box suitable for use in Class II locations. (Courtesy of Crouse-Hinds, Division of Cooper Industries)

> In locations where dust from magnesium, aluminum, aluminum bronze powders, or other metals of similarly hazardous characteristics may be present, fuses, switches, motor controllers, and circuit breakers shall have enclosures specifically approved for such locations.

Listed or approved dust-ignition proof enclosures are required for control transformers, solenoids, impedance coils, resistors, and any overcurrent devices or associated switching mechanisms (see Figure 5–6). This equipment shall not be installed in a location where dust from magnesium, aluminum, and other similar powders are present, unless they are installed within an enclosure approved specifically for the purpose.

Typical Explosionproof Electrical Controls.

Figure 5–6 Examples of enclosures suitable for use in accordance with Section 502–6(a)(2). (Courtesy of Crouse-Hinds, Division of Cooper Industries)

Aluminum bronze powders or other similarly hazardous metal dusts may not be present unless provided with an enclosure approved for the specific location. Motors, generators, and other rotating electrical machinery located in Division 1 locations shall be approved for the location or must be of a totally enclosed pipe-ventilated type in accordance with Sections 502–9 and 502–9(a) and must meet the temperature requirements in accordance with Section 502–1.

Lighting fixtures for fixed and portable lighting must be approved for the location and marked to indicate the maximum lamp wattage permitted. Where dust from magnesium, aluminum, aluminum bronze powders, or metals with similar characteristics may be present, the fixtures both fixed or portable and all auxiliary equipment must be approved for the location. Each fixture is to be protected from physical damage by a guard or be located where not subject to damage (see Figure 5–7).

Pendant and chain hung fixtures must be installed with an approved means and comply with the requirements of Section 502–11(a)(3) and supported in accordance with Section 502–10(a)(4).

> All utilization equipment *shall be* approved for Class II locations. In addition when dust from magnesium, aluminum, aluminum bronze powders, or other metals of similarly hazardous characteristics may be present, such equipment shall be approved for the specific location.

Flexible cords used in Class II Divisions 1 and 2 must comply with Section 502–12. They must be approved for extra-hard usage and supported by suitable means to prevent tension on the terminations, contain a grounding conductor complying with Section 400–23, and be spliced or terminated in an approved manner. Suitable seals shall be installed to prevent the entrance of dust where the flexible cord enters boxes or fittings that are required to be dust-ignition proof.

Receptacles and attachment plugs shall be designed, to provide connections for the grounding conductor of the flexible cord and shall be approved for Class II locations.

> (FPN): See *NEC*® Article 800 for rules governing the installation of communication circuits.

III. Equipment for Hazardous Areas

To make sure the safe operating temperatures of the fixtures will not be exceeded, maintenance personnel should always be sure to use the proper lamp specified by the manufacturer on the fixture nameplate.

Figure 5–7 Examples of lighting fixtures suitable for use in Class II locations. (Courtesy of Crouse-Hinds, Division of Cooper Industries)

Signaling, alarm, remote-control, and communication systems; and meters, instruments, and relays shall comply with the following:

The wiring methods shall be in accordance with Section 502–4(a). Enclosures approved for a Class II location are required for switches, circuit breakers, relays, contactors, fuses, and current-breaking contacts for bells, horns, howlers, sirens, and other devices that may produce sparks or arcs (see Figure 5–8). Resistors, transformers, and other heat-generating equipment shall be provided with enclosures approved for Class II locations. General-purpose enclosures are permitted where current-breaking contacts are immersed in oil or the interruption of current occurs within a chamber sealed against the entrance of dust.

Motors, generators, and other rotating electric machinery must meet the requirements of Section 502–8(a).

All wiring and equipment shall be approved for Class II locations associated with dusts that are electrically conductive. Where dust from magnesium, aluminum, aluminum bronze powders, or other metals of similarly hazardous characteristics may be present, all apparatus and equipment shall be approved for the specific conditions (see Figure 5–9).

> NEC® 502–15 states that "Live parts shall not be exposed."

DIVISION 1 GROUNDING REQUIREMENTS

Wiring and equipment must comply with the provisions of Article 250. Locknut-bushings and double-locknut types of contact are not acceptable for bonding purposes. Bonding jumpers or other approved means must be used to bond all intervening raceways, fittings, boxes, enclosures, and the like between Class II locations and the point of grounding for service equipment or separately derived systems, except as permitted in Section 502–16(a) and (b) and the exception. Where flexible conduit is permitted by Section 502–4, parallel internal or external bonding jumpers complying with Section 250–79 shall be installed.

> Additional applicable bonding requirements for hazardous (classified) locations are found in Section 250–78.

Surge arresters are designed for specific duty cycle and must comply with Article 280 and, in addition, shall be in suitable enclosures. Surge-protective capacitors must be designed for the specific duty in accordance with Article 460.

Signaling, Alarm, Remote-Control, and Communication Systems.

ETH Flex-Tone Signaling Device

ETW Telephone

EFS Pushbutton Station and pilot light

ESR Bell

ETH Horn signal

EV Strobe warning light

Figure 5–8 Examples of signaling, alarm, remote-control, and communications devices suitable for use in Class II locations. (Courtesy of Crouse-Hinds, Division of Cooper Industries)

Receptacle constructed with an interlocked switch. Rotating the plug after insertion actuates this switch. This is also referred to as "dead front."

Figure 5–9 ARE receptacle cutaway with a spring door. (Courtesy of Crouse-Hinds, Division of Cooper Industries)

A separate grounded conductor shall be installed in each single-phase branch circuit that is part of a multiwire branch circuit except where the disconnect device(s) opens all ungrounded conductors simultaneously.

CLASS II DIVISION 2 LOCATIONS

Enclosures for fuses, switches, circuit breakers, and motor controllers, including push buttons, relays, and similar devices, shall be dusttight (see Figure 5–9). Control transformers, solenoids, impedance coils, resistors, and resistance devices with associated equipment must be installed in dusttight enclosures. Control transformers, solenoids, and impedance coils without switching or arcing devices may be installed in tight metal housings without ventilating openings.

Electrically heated utilization equipment is generally required to be approved for Class II locations; see Section 502–10(b)(1) and exception. Motors and motor-driven utilization equipment are required to comply with Section 502–8(b) and exception.

Portable lighting equipment must be clearly marked to indicate maximum wattage and be approved for the location. Fixed lighting must be approved for Class II locations or shall provide enclosures for the lamps and lampholders that are designed to prevent the escape of sparks, burning material, or hot metal, and to minimize dust from accumulating on the lamps. Each fixture shall be marked with the maximum allowable wattage and protected from physical damage by suitable guards or by location. Pendant and chain hung fixtures shall comply with Section 502–10(b). Starting and control equipment for electric-discharge lamps shall comply with the requirements of Section 502–7(b).

Flexible cords in both Class II Division 1 and 2 are required to comply with Section 502–12. All flexible cords must be of a type approved for extra-hard usage with an equipment grounding conductor. Terminations must be made in an approved manner and the cord must be supported or fastened in a manner so that there will be no tension on the terminals. Receptacles and attachment plugs shall provide connections for the grounding conductor of the flexible cord and shall be designed so that connection to the supply circuit cannot be made or broken while live parts are exposed.

Signaling, alarm, remote-control, and communication systems; and meters, instruments, and relays shall comply with the following:

Arc-producing contacts shall be enclosed in approved enclosures in accordance with Section 502–14(a) or shall have tight metal enclosures designed to minimize the entrance of dust with no openings through which, after installation, sparks or burning material might escape. Nonincendive circuits that under normal conditions do not release sufficient energy to ignite a dust layer are permitted in general-purpose enclosures.

The windings and terminal connections of transformers and similar equipment shall be provided with tight metal enclosures without ventilating openings. Resistors and similar equipment shall comply with Section 502–14(a)(3). Where the maximum operating temperature will not exceed 120°C (248°F), general-purpose enclosures are permitted for thermionic tubes, nonadjustable resistors, or rectifiers.

Motors, generators, and other rotating electric machinery shall comply with Section 502–8(b) and the wiring method shall comply with Section 502–4(b).

NEC® Section 502–15 "Live parts shall not be exposed."

DIVISION 2 GROUNDING REQUIREMENTS

Wiring and equipment in Class II Division 2 locations shall be bonded as specified in Article 250 generally, Section 250–78, and Section 502–16(a) specifically. Flexible conduit used as an equipment-grounding conductor in accordance with Section 502–4 shall be installed with internal or external bonding jumpers complying with Section 250–79.

The bonding jumper may be deleted where listed liquidtight flexible metal is used and the conduit is 6 feet (1.83 m) or less in length, with listed fittings, having the overcurrent protection at 10 amperes or less, and the load is not a power utilization load.

Surge arresters, including their installation and connection, shall comply with Article 280. Surge-protective capacitors shall be of a type designed for specific duty.

WIRING METHODS FOR CLASS II LOCATIONS

Class II Division 1 Wiring Methods

Class II Division 1 locations permit threaded rigid metal conduit, threaded steel intermediate metal conduit, or type MI cable with termination fittings approved for the location and installed to avoid tensile stress at the fittings. The fittings and boxes with threaded bosses and close-fitting covers shall have no openings through which dust might enter or sparks or burning material might escape (see Figure 5–10). In locations where combustible dusts of an electrically conductive nature are likely, the fittings and boxes which are designed to contain taps, joints, or terminal connections must be approved for Class II locations. Flexible connections shall be made with dusttight flexible connectors, liquidtight flexible metal conduit with approved fittings, liquidtight flexible nonmetallic conduit with approved fittings, or flexible cord approved for extra-hard usage and provided with bushed fittings. Flexible cords must comply with Section 502–12. Flexible connections where subject to oil or other corrosive conditions shall be of a type approved for the condition or shall be protected by a suitable sheath.

> (FPN): See Section 502–16(b) for grounding requirements where flexible conduit is used.

Figure 5–10 An example of a Class II panelboard. (Courtesy of Crouse-Hinds, Division of Cooper Industries)

Class II Division 2 Wiring Methods

Class II Division 2 locations permit the use of rigid metal conduit, intermediate metal conduit, electrical metallic tubing, dusttight wireways, or types MI, MC, or SNM cable with approved termination fittings, or types MC, PLTC, or TC cable installed in ventilated channel-type cable trays in a single layer with a space of not less than the larger cable diameter between the two adjacent cables. Using any of the methods suitable for wiring in ordinary locations shall be permitted except wiring which under normal conditions cannot release sufficient energy to ignite a specific combustible dust mixture by opening, shorting or grounding.

> Nonincendive circuits are permitted in wiring methods suitable for ordinary locations.

SEALING REQUIREMENTS FOR CLASS II DIVISIONS 1 AND 2

Combustible and explosive dusts constitute a hazard in Class II locations. Such dusts do not travel for great distances within a conduit. Therefore, seals are not required if the conduit running between a dust-ignition proof enclosure and enclosure that is not dust-ignition proof consists of either a horizontal section not less than 10 feet in length or a vertical section not less than 5 feet in length extending downward from the dust-ignition proof enclosure through the raceway.

Where such conduit runs are not provided, seals must be used to prevent the entrance of dust into ignition-proof enclosures through the raceway. Sealing fittings and poured sealing compound may be used in both Division 1 and 2, and they are not required to meet the same stringent requirements as those in Section 501–5. The term "dust-ignition proof" is defined in *NEC®* Section 502–1.

REVIEW QUESTIONS

(Answers to some of these questions may be found in the introduction, previous chapters, and 1996 *NEC®*.)

1. Group _____ type dusts are the only Class II dusts that are considered electrically conductive.
   Answer: Reference:

2. Rooms or areas where dust removal equipment is installed and maintenance is assured may be _____ by the authority having jurisdiction.
   Answer: Reference:

3. Where a conduit is run horizontally for 30 feet before leaving the Class II Division 1 area, is a seal required to be installed within 1½ feet of the boundary, partition or wall?
   Answer: Reference:

4. May individual equipment be designed and listed and/or approved for both Class I and Class II locations?
   Answer: Reference:

5. May individual equipment be designed and listed and/or approved for both Division 1 and Division 2 locations?
   Answer: Reference:

6. Class I and Class II Division 1 equipment is designed to contain any explosion likely to occur within its enclosure.
   Class I only

   Class II only

   Both Class I and Class II

Neither Class I nor Class II

Answer: Reference:

7. Electrical metallic tubing is a permitted wiring method in a Class II Division 2 location.
   Answer: True or False.   Reference:

8. Surface temperature is an important consideration in Class I locations, but it is not a consideration in the listing and approval of equipment in Class II locations.
   Answer: True or False.   Reference:

9. Equipment and wiring of the type defined in Article 100 as explosion-proof shall not be required and shall not be acceptable in Class II locations unless approved for such locations.
   Answer: True or False.   Reference:

10. A circuit in which any spark or thermal effect is incapable of causing ignition of a mixture of flammable or combustible material in air under a prescribed test condition is a _____.
    Answer: Reference:

Chapter 6

Class III

CLASS III LOCATIONS

Class III locations are those that are hazardous because of the presence of easily ignitable fibers or flyings. It includes all locations in which the fibers or flyings are handled, manufactured, used, or stored. It is often believed that Class III hazards are far less hazardous than are Class I and Class II areas. This classification deals with different materials and a different type of hazard. However, they are not only dangerous because of ignitability but also because of propagation, the speed at which the fire will spread after it is ignited. When these fibrous materials, such as cotton lint that has settled on equipment, machinery, and building framing members catch on fire, the fire can and very often does spread at speeds approaching multiple explosions. These fires are called "flash fires," and have been the origins of some of our countries greatest disasters.

Class III Division 1 Locations

These locations are generally locations where the hazardous (classified) materials are manufactured, handled, and used.

Class III Location Examples

- Textile mills (i.e., cotton and rayon)
- Cotton gins
- Cotton seed mills
- Woodworking plants
- Combustible fiber manufacturing plants (i.e., hemp, cocoa, oakum, excelsior, cotton linters, and kapok)

(See Figures 6–1 and 6–2.)

Figure 6–1 Example of a Class III manufacturing area. (Courtesy of Appleton Electric Company)

Figure 6–2 Example of a Class III storage area. (Courtesy of Appleton Electric Company)

Class III Division 2 Locations

These locations are generally locations where easily ignitable fibers are stored or handled. Examples of locations where Division 2 areas may be are listed on previous page.

EQUIPMENT FOR CLASS III LOCATIONS

Class III locations require equipment which is designed to prevent the entrance of fibers and flyings, and to prevent the escape of sparks or burning material. Equipment installed in these locations must be capable of operating at its full load rating without developing surface temperatures hot enough to cause excessive dehydration or gradual carbonization of accumulated fibers or flyings, i.e., lint. Materials that are very susceptible to spontaneous ignition carbonize when they are excessively dry. Surface temperatures shall not exceed 165°C (329°F) for equipment not subject to overloading. Equipment, such as, motors or transformers, that may be overloaded are limited to a maximum operating temperature of 120°C (248°F). Transformers and capacitors located in Class III Divisions 1 and 2 must comply with the Class II requirements in Section 502–2(b).

SPECIFIC EQUIPMENT FOR CLASS III LOCATIONS

Class III Division 1 and 2 Locations

Dusttight enclosures are required for switches and motor controllers including push buttons, relays, similar devices, and overcurrent protection.

Dusttight enclosures meeting the temperature requirements in Section 503–1 are required for heat-producing equipment such as transformers and resistors used for the control of motors, appliances, and other electrical equipment.

Only totally enclosed nonventilated, totally enclosed pipe-ventilated, or totally enclosed fan-cooled motors, generators, and other rotating machinery are permitted unless otherwise allowed by the authority having jurisdiction. **Note:** Guidance to help the AHJ is found in Section 503–6 Exception.

Ventilating pipes shall comply with Section 503–7 and shall be constructed of metal 0.021 inch (533 micrometers) in minimum thickness, or of equally substantial noncombustible material, and shall be protected against physical damage and against corrosive influences. The piping shall be routed directly to a source of clean air outside with screened covers to prevent the entrance of small animals or birds.

Utilization equipment like electrically heated utilization equipment shall be approved for Class III locations. Motors of motor-driven utilization equipment must comply with Section 503–6. Switches, circuit breakers, motor controllers, and fuses must comply with Section 503–4.

Only enclosures for lamps and lampholders, which minimize the entrance of fibers and flyings and prevent the escape of sparks, burning material, or hot metal are permitted for fixed lighting. Fixtures shall not be subject to physical damage or shall be protected with suitable guards and be marked for the maximum permitted wattage. The surface temperature shall comply with Section 503–1 (see Figure 6–3). Pendant fixtures shall comply with Section 503–9(c). Portable lighting equipment shall comply with the provisions of Section 503–9(d).

Flexible cords shall be approved for extra-hard usage, contain a grounding conductor in accordance with Section 400–23, and shall be supported with an approved means to prevent tension on the terminations. Fittings shall be designed to prevent the entrance of fibers or flyings where the cord enters boxes, and all terminations shall be made in an approved manner.

Only receptacles and attachment plugs of the grounding type designed to minimize the accumulation or the entry of fibers or flyings and prevent the escape of sparks or molten particles are permitted unless otherwise permitted by the AHJ.

Signaling and intercommunication systems shall comply with the requirements of Article 503 regarding wiring methods, equipment, and related components.

Where installed for operation over combustible fibers or accumulations of flyings, traveling cranes and hoists for material handling, traveling cleaners for textile machinery, and similar equipment shall comply with the provisions of Section 503–13(a) through (d).

Storage-battery charging equipment shall be located in well-ventilated rooms to exclude flyings or lint and be built or lined with noncombustible materials.

Class III Divisions 1 and 2.

Live parts shall not be exposed, except as permitted in Section 503–13.

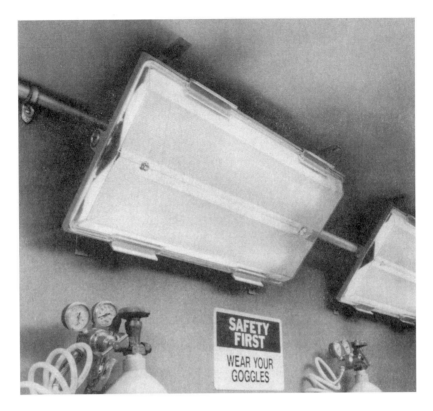

Figure 6–3 Example of a compact fluorescent light fixture suitable for use in Class III locations. (Courtesy of Appleton Electric Company)

WIRING METHODS

Class III Division 1 Locations

The permitted wiring methods are rigid metal conduit, intermediate metal conduit (IMC), electrical metallic tubing (EMT), rigid nonmetallic conduit, type MC, MI, or SNM cable with approved termination fittings or dusttight wireways. Flexible connections must be made with dusttight flexible fittings and installed in accordance with Section 503–16(b), using liquidtight flexible metal conduit with approved fittings, liquidtight flexible nonmetallic conduit with approved fittings, or flexible cord in conformance with Section 503–10 (see Figure 6–4).

Boxes and fittings must be dusttight.

Class III Division 2 Locations

Wiring methods approved for Class III Division 1 locations are acceptable. In addition, in compartments or locations solely for storage and containing no machinery, the use of open wiring on insulators may be used in accordance with all of the provisions of Article 320 when not subject to physical damage.

Grounding and Bonding Class III Divisions 1 and 2

> Class III wiring and equipment must comply with Article 250. See Section 503–16!

Bonding jumpers with proper fittings or other approved means of bonding shall be used and apply to all intervening raceways, fittings, boxes, enclosures, etc., between Class III locations and the point of grounding for service equipment or the point of grounding of a separately derived system except as otherwise permitted in Section 503–15 Exception.

> Locknut-bushing and double-locknut types shall not be depended upon for bonding, Section 503–16(a).

Equipment grounding conductors shall be installed in accordance with Section 250–91(b) and 503–3. Where flexible conduit is used, it must be installed with internal or external bonding jumpers in

Figure 6–4 Example of Class III conduit bodies. (Courtesy of Crouse-Hinds, Division of Cooper Industries)

parallel with each conduit and complying with Section 250–79 or in accordance with Section 503–15(b). Exception where all of the conditions are met.

SEALING REQUIREMENTS

Seals are not required in Class III locations.

REVIEW QUESTIONS

(Answers to some of these questions may be found in the introduction, previous chapters, and 1996 *NEC®*.)

1. Motors approved for Class II Division 1 shall be permitted for use in Class III Division 1 and 2 locations.
   Answer: True or False.   Reference:

2. Flexible cords permitted in Class III locations shall be approved for _____ usage as permitted in Article _____.
   Answer: Reference:

3. The surface temperature requirements in Article 503 for equipment are primarily concerned with _____ combustion.
   Answer: Reference:

4. When are seals required in Class III Division 1 locations?
   Answer: Reference:

5. Bonding conductors are not required where they do not exceed _____ feet in length and the enclosed circuits are of _____ amperes or less.
   Answer: Reference:

6. Electrical metallic tubing and type AC cable are approved for Class III Division 1 locations.
   Answer: True or False.   Reference:

7. Live parts are not permitted to be exposed generally. However, _____ _____, _____, and _____ are permitted to have exposed live parts under special conditions.
   Answer: Reference:

8. Double-locknut and locknut-bushing type connections are not permitted in Class III Division 1 locations, but are permitted in Class III Division 2 locations.
   Answer: True or False.   Reference:

9. Ventilating piping in Class III Division 1 and 2 is permitted to be _____ or _____ material, and outside openings shall be covered with _____.
   Answer: Reference:

10. Name four types of materials associated with easily ignitable fibers and flyings that may be found in a Class III location.
    Answer: Reference:

Chapter 7

Hazardous (Classified) Locations Specific

Article 510 states that Articles 511 through 517 cover occupancies or parts of occupancies that are or may be hazardous because of atmospheric concentrations of gases, vapors, liquids, or materials that may be readily ignitable.

Electrical equipment and electrical wiring in occupancies within the scope of Articles 511 through 517 apply to the general rules of Chapters 1 through 4 except as modified by these articles. Where judged that unusual conditions exist, the authority having jurisdiction has the authority and may rule as to the application of the specific rules.

COMMERCIAL GARAGES, REPAIR AND STORAGE

> **Scope:** Article 511, Section 511–1, "These occupancies shall include locations used for service and repair operations in connection with self-propelled vehicles (including passenger automobiles, buses, trucks, tractors, etc.) in which volatile flammable liquids are used for fuel or power." (See Figure 7–1.)

Commercial garages are defined by *NEC*® Article 511. Additional information may be obtained in NFPA 88A–1991 for parking structures and NFPA 88B–1991 for repair garages. Although Article 511 only covers those areas within the scope of NFPA 88A and 88B, it may provide good guidelines for similar areas in which other combustible engines are being serviced and repaired. It is important to bear in mind that similar commercial repair shops for such equipment as boat

Figure 7–1 Lubrication pit area for a commercial garage. (Courtesy of Appleton Electric Company)

repair shops, lawnmower repair shops, and other shops used to store or repair volatile fuel-driven or propelled equipment, although similar in nature is not within the scope of Article 511. These gasoline engine repair shops or areas within these shops may be hazardous (classified) locations and as such must comply with the appropriate provisions of Article 501.

> Areas in which flammable fuel is transferred to vehicle fuel tanks must conform with *NEC®* Article 514.

Parking garages, which are used for parking or storage and where no repair work is done except the exchange of parts and routine maintenance requiring no use of electrical equipment, open flame, welding, or the use of volatile flammable liquids, are not required to be classified, but they shall be required to be adequately vented with the capacity to carry the exhaust fumes of the engines.

Article 500 Areas within 511 Structures

Class I Division 2 shall apply to areas up to a level of 18 inches above each floor. The entire area up to a level of 18 inches above the floor shall be considered that classified location.

> Except where the enforcing agency judges there is mechanical ventilation providing a minimum of four air changes per hour.

Class I Division 1 shall apply to any pit or depression below floor level and shall apply to the area up to floor level. Where ventilation is provided in which six air changes per hour are exhausted at the floor level of the pit, the area may be classified as Class I Division 2 by the enforcing agency (see Figure 7–2).

A change to the 1996 *NEC®* brought about a special set of rules for lubrication and service rooms. This was because of the limited amounts gasoline and cleaning solvents present in these normally franchised facilities. Although they clearly are commercial

> Section 514–3(b) Exception "Lubrication and service rooms (e.g., Quickie Lube franchises) without dispensing shall be classified in accordance with Table 514–2."

> TABLE 514–2 (Partial Extract to Cover This Specific Requirement)
>
> Lubrication or Service Room—
>
> Class 1 Division 2
>
> Entire area within any pit used without dispensing for lubrication or similar services where Class I liquids may be released.
>   Area up to 18 inches above any such pit, and extending a distance of 3 feet horizontally from any edge of the pit.
>   Entire unventilated area within any pit, below-grade area, or subfloor area.
>   Area up to 18 inches above any such unventilated pit, below-grade work area, or subfloor work area and extending a distance of 3 feet horizontally from the edge of any such pit, below-grade work area, or subfloor work area.
>
> Nonclassified
>
> Any pit, below-grade work area, classified or subfloor work area that is ventilated in accordance with 5–1.3. NFPA 30A 1990

garages, a new exception was added to Section 511–3(b) which assigns these facilities to Article 514.

Areas not classified are areas separated and adjacent to classified locations. They are specifically defined and are areas or rooms in which flammable vapors are not likely to be released or to accumulate, i.e., stock rooms, switchboard rooms, and similar locations, where mechanically ventilated at a rate of four or more air changes per hour or where effectively cut off by walls or partitions.

The authority having jurisdiction may deem that in specific adjacent areas where no ignition hazard exists, the area may not require classification. The adequate ventilation, air pressure differentials, or physical spacing are valid reasons for making this determination.

Figure 7–2 Installation guidelines for a commercial garage area. (Courtesy of Appleton Electric Company)

> Where volatile fuel-dispensing units are located within buildings, the requirements of Article 514 apply. Where mechanical ventilation is provided in the dispensing area, the controls shall be interlocked so that the fuel dispenser cannot operate without ventilation as prescribed in Section 500–5(b).

> Liquid petroleum gas dispensers are prohibited within Article 511 structures.

All portable lighting equipment shall be approved for Class I Division 1 unless the lamp and its cord are supported or arranged in such a manner that they cannot be used in the Section 511–3 locations (see Figure 7–3).

Wiring and Equipment

Wiring and equipment located within the Class I locations shall conform to the applicable provisions of Article 501. For raceways imbedded in masonry walls or buried beneath the floor and if any connections or extensions lead into or through the area, seals are required as per Section 501–5 and 501–5(b)(2) and

Figure 7–3 Example of a portable flood light suitable for use in hazardous locations; for other examples of portable lighting, refer to Figure 4–9 in Chapter 4. (Courtesy of Appleton Electric Company)

shall apply to both horizontal and vertical boundaries of the defined Class I location (see Figures 7–2, 7–4, and 7–5).

> **Exception No. 2:** Rigid nonmetallic conduit complying with Article 347 shall be permitted where buried under not less than 2 feet (610 mm) of cover. Where rigid nonmetallic conduit is used, threaded rigid metal conduit or threaded steel intermediate metal conduit shall be used for the last 2 feet (610 mm) of the underground run to emergence or to the point of connection to the aboveground raceway and an equipment grounding conductor shall be included to provide electrical continuity of the raceway system and for grounding of the noncurrent carrying metal parts.

Sealing Requirements

Approved seals are required in commercial garages just as they are for any Class I Division 1 or 2 area. The seal fittings must be installed in accordance with the provisions of Section 501–5 for horizontal as well as vertical boundaries of the defined locations (see Figure 7–5).

Wiring above Class I Locations

The wiring above the Class I location must be installed in metal raceways, nonmetallic conduit, electrical nonmetallic tubing, flexible metal conduit, liquidtight flexible metal conduit, liquidtight flexible nonmetallic conduit, or shall be type MC, MI manufactured wiring systems, SNM, or PLTC cable in accordance with Article 725. TC cable, cellular metal floor raceway or cellular concrete floor raceways are permitted to be used below the floor, but such raceways shall have no connections leading into or through the Class I location above the floor (see Figure 7–2).

Where a circuit is portable or supplied from pennants and includes a grounded conductor as provided in Article 200, receptacles, attachment plugs, connectors, and similar devices shall be of the polarized type and the grounded conductor of the flexible cord shall be connected to the screw shell of the lampholder or to the grounded terminal of any utilization equipment.

Figure 7–4 MI cable installation (Article 330). (Courtesy of Pyrotenax USA Inc., photo by Robert Stewart Associates)

Figure 7–5 Seal fitting installation guidelines and examples of fittings. (Courtesy of Appleton Electric Company)

Equipment Located above Class I Locations

Where equipment is located above the Class I location, the equipment shall be at least 12 feet above the floor level. Where it may be deemed as arc-producing equipment, which may produce arcs, sparks, or particles of hot metal such as cut-out switches, charging panels, generators, motors, or other similar equipment. Having make-and-break sliding contacts, it is required to be of the totally enclosed type or so constructed as to prevent the escape of sparks or hot metal particles (see Figure 7–2). This requirement does not include receptacles, lamps, or lampholders.

Fixed lighting as located over the lanes through which vehicles are commonly driven or otherwise exposed to physical damage shall be located not less than 12 feet above the floor unless totally enclosed or so constructed as to prevent the escape of arcs, sparks, or hot metal particles.

Battery chargers and their control equipment are not permitted to be located within the classified locations in Section 511–3.

Ground-Fault Circuit-Interrupter Protection

Section 511–10 requires that all 125-volt single-phase 15- and 20-ampere receptacles installed in areas where electrical diagnostic equipment, electrical hand tools, or portable lighting equipment are to be used shall have ground-fault circuit-interrupter protection for personnel.

Although this article only covers specific areas, it may provide good guidelines for similar areas in which other combustible engines are being serviced and repaired. Bear in mind that boat shops, lawnmower shops, and the like, although similar in nature, must comply with Article 501 in areas classified as such.

Grounding

In addition to the general requirements of Article 250 which covers all areas in a commercial garage, the classified areas shall be grounded in accordance with Section 501–16.

REVIEW QUESTIONS

(Answers to some of these questions may be found in the introduction, previous chapters, and 1996 *NEC®*.)

1. Portable lighting equipment must be Class _____ Division _____ except where supported and arranged in a manner where this equipment cannot be used in Section 511-3 areas.
   Answer: True or False.   Reference:

2. Nonmetallic sheathed cable can only be used in areas above the classified areas.
   Answer: True or False.   Reference:

3. Installations of lubritoriums and service rooms must meet the requirements of Article _____.
   Answer: Reference:

4. Article 511 covers all commercial repair garages and parking structures including one- and two-family dwellings.
   Answer: True or False.   Reference:

5. The classified hazardous locations within a snowmobile repair shop are required to be wired in accordance with Article _____.
   Answer: Reference:

6. The authority having jurisdiction may _____ areas with adequate ventilation.
   Answer: Reference:

7. Is a GFCI ground-fault circuit interrupter required for a receptacle located behind a soft drink vending machine in a commercial garage?
   Answer: Reference:

8. Which areas in a commercial garage would likely be classified Class II Division 2?
   Answer: Reference:

9. Which areas in a commercial garage are likely to be classified Class I Division 1?
   Answer: Reference:

10. Most public parking garages are classified as _____ locations.
    Answer: Reference:

Chapter 8

Aircraft Hangars

Aircraft hangars are defined by *NEC®* Article 513 as locations used for storage or servicing of aircraft in which gasoline, jet fuels, or other volatile, flammable liquids or flammable gases are used (see Figure 8–1). It shall not include locations used exclusively for aircraft that have never contained such liquids or gases or that have been drained and properly purged. This would include areas such as assembly plants and rebuild plants for aircraft. Although it is getting to be increasingly popular for residential developments to be built in which individual aircraft are stored adjacent to the residence, this article has not been revised to cover those areas, and therefore where encountered, they must meet the provisions of *NEC* Article 513. Applicable building codes may also have additional requirements related to these special residential developments with an airstrip and individual aircraft hangar on each resident's property.

Figure 8–1 Recommended installation guidelines for aircraft hangars. (Courtesy of Appleton Electric Company)

Specific designated areas within each aircraft hangar are classified according to the degree of hazard that may exist generally. The authority having jurisdiction has the final determination regarding the classification of locations. However, the requirements of the *NEC®* are the minimal requirements and are generally applicable.

Class I Division 1 Areas

Any pit or depression below the level of the hangar floor extending up to floor level.

Class I Division 2 Areas

The entire area of the hangar, including any adjacent and communicating areas not suitably cut off from the hangar up to a level 18 inches (457 mm) above the floor. The area within 5 feet (1.52 m) horizontally from aircraft power plants (engines) and/or aircraft fuel tanks and shall extend upward from the floor to a level 5 feet (1.52 m) above the upper surface of wings and/or engine enclosures (see Figure 8–2).

Nonclassified Areas

Any areas that are suitably cut off from the classified portion of the hangar by walls or partitions and are ventilated so that flammable liquids or vapors are not likely to be present (i.e., stockrooms, electrical control rooms, and other similar locations) need not be classified.

Wiring and Equipment in Classified Locations

All wiring and equipment that is or may be installed and/or operated within hangar areas that are classified Class I Divisions 1 or 2, shall comply with the applicable provisions of Article 501. Wiring installed in or under the hangar floor shall comply with the requirements for Class I Division 1. Adequate drainage is required where classified wiring is located in vaults, pits, or ducts. The wiring shall not be installed in enclosures or compartments with any service other than piped compressed air. Attachment plugs and receptacles in classified locations shall be of a type approved for the location in accordance with Article 501 (see Figure 8–3). Any jet-fueling station that is normally located near the aircraft, the hangars, and runways must comply with Articles 514 and 515.

Wiring and Equipment Not within Class I Locations

All fixed wiring in a hangar, but not within a classified location, shall be installed in metal raceways or shall be type MI, TC, SNM, or type MC cable (refer to Figure 4–17 in Chapter 4 and Figure 7–4 in Chapter 7).

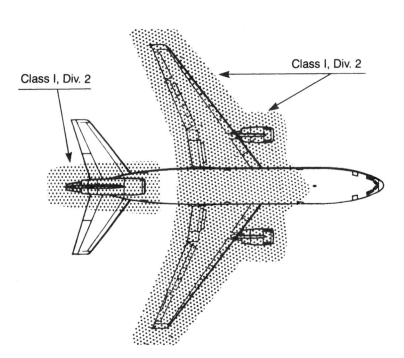

Figure 8–2 Top view of Class I Division 2 areas. (Courtesy of Appleton Electric Company)

Aircraft Hangars

Figure 8–3 Typical jet fuel bulk fuel station. (Courtesy of Appleton Electric Company)

Wiring suitably cut off from the hangar area as defined in Section 513–2(d) may be of the other suitable wiring methods in *NEC®* Chapter 3.

Where portable equipment, lamps, or pendants are installed, they must be of a flexible cord suitable for the type of service, and the cord must contain a separate green-equipment grounding conductor.

For circuits supplying portable equipment or pendants which include a grounded conductor as provided in Article 200, the grounded conductor is to be connected to the screw shell of lampholders and to the grounded terminal of all utilization equipment supplied. All receptacles, attachment plugs, connectors, and similar devices must be the grounding type. An approved means must be provided for maintaining continuity of the grounding conductor between the fixed wiring system and the noncurrent-carrying metal portions of pendant fixtures, portable lamps, and portable utilization equipment.

Equipment that is less than 10 feet above wings and/or engine enclosures of aircraft but is not within the classified area as described in Section 513–2(a) through (c) of the *NEC®*, that may produce arcs, sparks, or particles of hot metal, such as lampholders for fixed lighting, switches, receptacles, or other equipment having make-and-break or sliding contacts, must be of the totally enclosed type or designed to prevent escape of sparks or hot metal particles. General-purpose type equipment may be used in those areas described in Section 513–2(d); however, some restrictions apply to the specific type of lampholders and portable equipment that may be used.

Stanchions, Rostrums, and Docks

All electric wiring, outlets, and equipment (including lamps) on or attached to stanchions, rostrums, or docks that are located or likely to be located in a Class I location as defined in Section 513–2(c) must comply with the requirements for Class I Division 2 locations. Where they are not located and are not likely to be located in a Class I location as defined in Section 513–2(c), wiring and equipment located more than 18 inches above the floor need only comply with Sections 513–4 and 513–5. Wiring and equipment not more than 18 inches above the floor in any position shall comply with the requirements of Section 513–5(a), and all receptacles and attachment plugs must be of the locking type. Mobile stanchions not intended to be used within the classified areas are required to have at least one permanently affixed warning sign to read:

> "**Warning**: Keep 5 feet clear of aircraft engines and fuel tank areas."

SEALING

The rules for sealing in aircraft hangars are defined by Class and Division (refer to Figure 7–5 in Chapter 7). Seals must be installed in accordance with the provisions of Section 501–5. These requirements apply to horizontal as well as to vertical boundaries of the defined Class I locations. Raceways embedded in a concrete floor or buried beneath a floor are considered to be within the Class I location.

Additional Specific Requirements

Section 513–8 requires that all aircraft electrical systems be de-energized when the aircraft is stored in a hangar. This section recommends that systems be de-energized during maintenance operations whenever possible.

> Aircraft batteries shall not be charged where installed in an aircraft located inside or partially inside a hangar. Battery chargers and their control equipment are not permitted to be located or operated within any of the Class I locations defined in Section 513–2.
>
> Tables, racks, trays, and wiring are not permitted within a Class I location and, in addition, must comply with Article 480 for storage batteries.

It is recommended that charging equipment be located in a separate building or in a nonclassified area such as defined in Section 513–2(d). All mobile chargers are required to have at least one permanently affixed warning sign to read:

> "**Warning**: Keep 5 feet clear of aircraft engines and fuel tank areas."

Aircraft energizers (external power sources) shall be so designed that all electric equipment and fixed wiring will be at least 18 inches above floor level and shall not be operated in a Class I location as defined in Section 513–2(c). Mobile energizers must have at least one permanently affixed warning sign to read:

> "**Warning**: Keep 5 feet clear of aircraft engines and fuel tank areas."

Mobile servicing equipment (such as vacuum cleaners, air compressors, and air movers) having electric wiring and equipment not suitable for Class I Division 2 locations must be so designed that all fixed wiring and equipment will be at least 18 inches above the floor. This equipment is not permitted to be operated within the Class I location as defined in Section 513–2(c) and must have at least one permanently affixed warning sign to read:

> "**Warning**: Keep 5 feet clear of aircraft engines and fuel tank areas."

Equipment that is not identified as suitable for Class I Division 2 locations is not permitted to be operated in locations where, during maintenance operations, flammable liquids or vapors are likely to be released.

Grounding

All metal raceways, metal-jacketed cables, and all noncurrent-carrying metal portions of fixed or portable equipment, regardless of voltage, shall be grounded as provided in Article 250.

REVIEW QUESTIONS

(Answers to some of these questions may be found in the introduction, previous chapters, and 1996 *NEC®*.)

1. In Article 513 of the *NEC®*, the area within 5 feet of the wings or engine enclosures of aircraft being assembled is considered to be classified _____.
   Answer: Reference:

2. May battery chargers be connected directly to the battery in an aircraft within the aircraft hangar?
   Answer: Reference:

3. Are seals required in horizontal runs of conduit at least 10 feet in length in the Division 1 areas of aircraft hangars?
   Answer: Reference:

4. General-purpose equipment, such as vacuum cleaners and air moving equipment, can only be used near equipment in a _____ area.
   Answer: Reference:

5. Depressions in the floor or pits must be classified as a Class _____ Division _____ area.
   Answer: Reference:

6. Wiring methods of a type as listed in Section _____ are required and must be used in all Class I Division 1 locations within the aircraft hangar.
   Answer: Reference:

7. Where would the definition of an aircraft hangar as it applies to the code be found?
   Answer: Reference:

8. All portable equipment that may be used within aircraft hangars must be approved for Class _____ Division _____ locations.
   Answer: Reference:

9. Toolrooms and restrooms may not be classified when suitably cut off by walls or partitions or adequately vented.
   Answer: True or False.   Reference:

10. All portable lighting equipment used with aircraft hangars must be suitable for _____ .
    Answer: Reference:

Chapter 9

Gasoline Dispensing and Service Stations

Gasoline dispensing and service stations are defined in *NEC®* Article 514 as a location where gasoline or other volatile, flammable liquids or liquefied flammable gases are transferred to fuel tanks, including auxiliary fuel tanks of self-propelled vehicles or approved containers. These are areas where flammable liquids having a flash point below 38°C (100°F), such as gasoline or liquid petroleum gas, are dispensed. These are generally areas within a facility and not the entire facility. For example, a truck stop or combination use facility may contain several type areas other than a gasoline dispensing and service station. It may contain a place of assembly as covered in Article 518, like a restaurant. It may also contain an area that is classified as a commercial garage as covered in Article 511, and if suitably separated, it may contain a diesel fuel-dispensing area that is not required to comply with the provisions of this article. Resorts commonly have more than one area coming under the provisions of this article, such as a gasoline-dispensing island for retail sales and one for their own equipment, a propane tank fill area, and a marine area for fueling boats. Other areas, such as lubritoriums, service rooms, repair rooms, offices, sales rooms, compressor rooms, and similar locations, shall comply with *NEC®* Articles 510 and 511 with respect to the electrical wiring and equipment. Where the authority having jurisdiction can satisfactorily determine that flammable liquids having a flash point below 38°C (100°F), such as gasoline, will not be handled, such locations need not be classified. Additional information regarding the safeguards for gasoline dispensing and service stations may be obtained by seeing the Automotive and Marine Service Station Code NFPA 30. For additional information pertaining to LP gas systems, other than residential or commercial, see NFPA 58. For storage and handling of liquefied petroleum gases, see NFPA 59. For gasoline-dispensing stations and marine applications, such as boatyards, see *NEC®* Section 555–9. It is important that each area be classified separately to assure a safe installation that will remain safe throughout the life of the facility with proper maintenance.

CLASS I LOCATIONS

The areas within the scope of Article 514 that are to be classified Class I are clearly defined in Table 514–2, and the boundaries for each area to be classified does not extend beyond the unpierced wall, the roof, or other solid partitions of that area. In open air areas, the classified area boundaries are measured in feet or inches from the hazardous substance source, such as the fill opening, dispenser, or vent discharge. The likelihood of the presence of the liquid or vapors dictates the division. The group for these materials is Group D. All wiring and electrical equipment covered within these classified Class I locations are required to

Figure 9–1 Example of a typical gasoline island area. (Courtesy of Appleton Electric Company)

comply with the applicable provisions of Article 501 except where other wiring methods are specifically allowed by Section 514–8.

The space within a dispenser up to a height of 4 feet and the space within 18 inches horizontally of a dispenser and extending to a height 4 feet above its base must be classified as a Class I Division 1 location (see Figure 9–1).

> Space classification inside the dispenser enclosure is covered in *Power Operated Dispensing Devices for Petroleum Products,* ANSI/UL87–1990.

Any wiring or equipment that is installed beneath any part of a Class I Division 1 or 2 location is classified as being within the Class I Division 1 location to the point of emergence from below grade. Where the dispensing unit or its hose or nozzle valve hasn't been suspended from overhead, the space within the dispenser enclosure and the area within 18 inches in all directions from the enclosure that is not suitably cut off by a ceiling or wall, is to be classified as a Class I Division 1 area.

All equipment integral with the overhead dispensing hose and nozzle must be suitable for Class I Division 1 hazardous locations. A spherical volume within a 3-foot radius of the point of discharge of any tank vent pipe is classified as a Class I Division 1 location.

> Lubritoriums, service rooms, and repair rooms, as well as other areas not suitably cut off or elevated at least 18 inches above these areas, are classified and considered as commercial garages. These areas generally contain areas that are classified as Class I Division 1 and except for pits and depressions in floors, many of these areas, when properly designed, can be reduced to Division 2 or nonclassified.

CLASS I DIVISION 2

The area within 2 feet in all directions from the Class I Division 1 area and extending down to grade level is classified as a Class I Division 2 area. The horizontal area 18 inches above grade and extending 20 feet measured from those points vertically below the outer edges of the overhead dispensing enclosure are also classified as a Class I Division 2 area (see Figures 9–2 and 9–3).

The area within 3 feet measured in any direction from the dispensing point of a hand-operated unit dispensing a Class I flammable liquid, such as white gas, is considered a Class I Division 2 location. An outside location in buildings not suitably cut off, any area not classified as a Class I Division 1 location that is within 20 feet horizontally of the interior enclosure of the dispensing pump or within 10 feet horizontally from a tank filled by and less than 18 inches above grade, is to be classified as a Class I Division 2 location (see Figure 9–1). That part of the spherical volume lying between a 3-foot radius and a 5-foot radius of the point

Table 514–2.* Class I Locations—Service Stations

Location	Class I, Group D Division	Extent of Classified Location
Underground Tank		
Fill Opening	1	Any pit, box, or space below grade level, any part of which is within the Division 1 or 2 classified location.
	2	Up to 18 inches above grade level within a horizontal radius of 10 feet from a loose fill connection and within a horizontal radius of 5 feet from a tight fill connection.
Vent—Discharging	1	Within 3 feet of open end of vent, extending in all directions.
Upward	2	Space between 3 feet and 5 feet of open end of vent, extending in all directions.
Dispensing Device[1†]		
(except overhead type)*		
Pits	1	Any pit, box, or space below grade level, any part of which is within the Division 1 or 2 classified location.
Dispenser		(FPN): Space classification inside the dispenser enclosure is covered in *Power Operated Dispensing Devices for Petroleum Products,* ANSI/UL 87–1990.
	2	Within 18 inches horizontally in all directions extending to grade from (1) the dispenser enclosure or (2) that portion of the dispenser enclosure containing liquid handling components.
		(FPN): Space classification inside the dispenser enclosure is covered in *Power Operated Dispensing Devices for Petroleum Products,* ANSI/UL 87–1990.
Outdoor	2	Up to 18 inches above grade level within 20 feet horizontally of any edge of enclosure.
Indoor		
with Mechanical Ventilation	2	Up to 18 inches above grade or floor level within 20 feet horizontally of any edge of enclosure.
with Gravity Ventilation	2	Up to 18 inches above grade or floor level within 25 feet horizontally of any edge of enclosure.
Dispensing Device†		
Overhead Type*	1	The space within the dispenser enclosure, and all electrical equipment integral with the dispensing hose or nozzle.
	2	A space extending 18 inches horizontally in all directions beyond the enclosure and extending to grade.
	2	Up to 18 inches above grade level within 20 feet horizontally measured from a point vertically below the edge of any dispenser enclosure.
Remote Pump—Outdoor	1	Any pit, box, or space below grade level if any part is within a horizontal distance of 10 feet from any edge of pump.
	2	Within 3 feet of any edge of pump, extending in all directions. Also up to 18 inches above grade level within 25 feet horizontally from any edge of pump.
Remote Pump—Indoor	1	Entire space within any pit.
	2	Within 5 feet of any edge of pump, extending in all directions. Also up to 3 feet above grade level within 25 feet horizontally from any edge of pump.

Figure 9–2 Table 514–2. (Reprinted with permission from NFPA 70–1996, *National Electrical Code*®, Copyright © 1995, National Fire Protection Association, Quincy, Massachusetts 02269. This reprinted material is not the complete and official position of the NFPA on the referenced subject, which is represented only by the standard in its entirety)

Table 514–2.* Class I Locations—Service Stations (continued)

Location	Class I, Group D Division	Extent of Classified Location
Lubrication or Service Room— with Dispensing	1	Any pit within any unventilated space.
	2	Any pit with ventilation.
	2	Space up to 18 inches above floor or grade level and 3 feet horizontally from a lubrication pit.
Dispenser for Class I Liquids	2	Within 3 feet of any fill or dispensing point, extending in all directions.
Lubrication or Service Room— without Dispensing	2	Entire area within any pit used for lubrication or similar services where Class I liquids may be released.
	2	Area up to 18 inches above any such pit, and extending a distance of 3 feet horizontally from any edge of the pit.
	2	Entire unventilated area within any pit, below grade area, or subfloor area.
	2	Area up to 18 inches above any such unventilated pit, below grade work area, or subfloor work area and extending a distance of 3 feet horizontally from the edge of any such pit, below-grade work area, or subfloor work area.
	Non-classified	Any pit, below-grade work area, or subfloor work area that is provided with exhaust ventilation at a rate of not less than 1 cfm per sq ft (0.3 cu m per min per sq m) of floor area at all times that the building is occupied or when vehicles are parked in or over this area and where exhaust air is taken from a point within 12 inches (0.3 m) of the floor of the pit, below-grade work area, or subfloor work area.
Special Enclosure Inside Building**	1	Entire enclosure.
Sales, Storage, and Rest Rooms	Non-classified	If there is any opening to these rooms within the extent of a Division 1 location, the entire room shall be classified as Division 1.
Vapor Processing Systems Pits	1	Any pit, box, or space below grade level, any part of which is within a Division 1 or 2 classified location or that houses any equipment used to transport or process vapors.
Vapor Processing Equipment Located within Protective Enclosures (FPN): See *Automotive and Marine Service Station Code*, NFPA 30A-1993, Section 4-5.7	2	Within any protective enclosure housing vapor processing equipment.
Vapor Processing Equipment Not within Protective Enclosures (excluding piping and combustion devices)	2	The space within 18 inches in all directions of equipment containing flammable vapor or liquid extending to grade level. Up to 18 inches above grade level within 10 feet horizontally of the vapor processing equipment.
Equipment Enclosures	1	Any space within the enclosure where vapor or liquid is present under normal operating conditions.
Vacuum-Assist Blowers	2	The space within 18 inches in all directions extending to grade level. Up to 18 inches above grade level within 10 feet horizontally. For SI units: 1 inch = 2.5 cm; 1 feet = 0.3048 m.

[1]Refer to Figure 1 for an illustration of classified location around dispensing devices.
*Ceiling mounted hose reel.
**(FPN): See *Automotive and Marine Service Station Code*, NFPA 30A–1993, Section 2-2.
[†](FPN): Area classification inside the dispenser enclosure is covered in *Power-Operated Dispensing Devices for Petroleum Products*, ANSI/UL 87–1990.

Figure 9–2 *(continued)* Table 514–2. (Reprinted with permission from NFPA 70–1996, *National Electrical Code*®, Copyright © 1995, National Fire Protection Association, Quincy, Massachusetts 02269. This reprinted material is not the complete and official position of the NFPA on the referenced subject, which is represented only by the standard in its entirety)

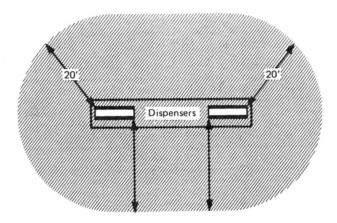

Figure 1. Classified Locations Adjacent to Dispensers as Detailed in Table 514-2.

Exception No. 1: Type MI cable shall be permitted where it is installed in accordance with Article 330.

Exception No. 2: Rigid nonmetallic conduit complying with Article 347 shall be permitted where buried under not less than 2 feet (610 mm) of earth. Where rigid nonmetallic conduit is used, threaded rigid metal conduit or threaded steel intermediate metal conduit shall be used for the last 2 feet (610 mm) of the underground run to emergence or to the point of connection to the aboveground raceway; an equipment grounding conductor shall be included to provide electrical continuity of the raceway system and for grounding of noncurrent-carrying metal parts.

Figure 9–3 Figure 1 detailed to Table 514–2. (Reprinted with permission from NFPA 70–1996, *National Electrical Code*®, Copyright © 1995, National Fire Protection Association, Quincy, Massachusetts 02269. This reprinted material is not the complete and official position of the NFPA on the referenced subject, which is represented only by the standard in its entirety)

of discharge of any tank vent pipe is classified as a Class I Division 2 location. If the tank vent does not discharge upward, the area extending to grade level is classified as a Class I Division 2 location.

AREAS NOT CLASSIFIED

The space beyond the 5-foot radius from the tank vents that discharge upward and spaces beyond unpierced walls and areas below grade that lie beneath tank vents are not classified as hazardous.

Seals

> **Warning:** Unless specifically marked, seals are limited to a 25% wire fill. Generally the code will permit an oversized seal to achieve the 40% wire fill allowed for the conduit. Some manufacturers are now producing a seal that will permit a 40% wire fill.

A seal is required as the first fitting after a conduit emerges from the earth or concrete and enters a dispenser or any cavity or enclosure in connection with it. A seal is required in any conduit that extends upward beyond the vertical boundary of a Class I Division 2 area, such as a conduit supplying a lighting standard within a hazardous area (refer to Figure 7–5 in Chapter 7). The required seal must be placed so that no coupling or fitting of any kind is between the seal and the boundary line dividing the hazardous and nonhazardous areas. A seal is required at or beyond the horizontal boundary at the point where the conduit emerges from below grade. The seal must be the first fitting where the conduit emerges into a nonhazardous area. If the conduit does not extend beyond the 20-foot horizontal boundary, but emerges and extends upward while still within the hazardous area, the seal must be so placed that no coupling or fitting is between it and the vertical boundary. Circuits leading to or through a gasoline dispenser enclosure (see Figure 9–1) must be provided with switches or circuit breakers which simultaneously disconnect all circuit conductors, including the neutral, from the source of supply. Section 514–5 states that single pole breakers utilizing handle ties are not permitted for this application.

Circuit Disconnects

A disconnecting means is required for each circuit leading to or through all dispensing equipment. The switch or other acceptable means must disconnect simultaneously all conductors, including the grounded conductor. Specific rules for stations that have attendants on duty and those that are unattended as to the type and location of required emergency controls can be found in NFPA 30A. Extracts from this standard can be found in Section 514–5(b) and (c).

REVIEW QUESTIONS

(Answers to some of these questions may be found in the introduction, previous chapters, and 1996 *NEC*®.)

1. Does Article 514 apply to the underground wiring to a sign that is at least 30 feet from any hazardous location in a service station?
   Answer: Reference:

2. A diesel fuel dispenser is mounted adjacent to a gasoline dispenser on a service station island. Is a sealing fitting required in the conduit entering the diesel fuel dispenser?
   Answer: Reference:

3. Can intermediate metal conduit be used in a Class I Division 1 location?
   Answer: Reference:

4. Can rigid nonmetallic conduit be used in a Class I Division 1 location?
   Answer: Reference:

5. In a highly hazardous location where propane is used, the customer states that the wiring to the control circuit is of sufficient low voltage that a spark cannot be generated from this circuit. The devices and wiring have a control drawing number on them. What type of wiring is most likely employed in this installation where there is not enough current or voltage to generate a spark sufficient to cause an explosion?
   Answer: Reference:

6. In question 5, how many conduit seals are required?
   Answer: Reference:

7. A seal must be placed within _____ inches of the boundary where a conduit leaves a Class I Division 1 location and enters a Class I Division 2 location.
   Answer: Reference:

8. Where in the *NEC®* does it prohibit the use of handle ties on single pole circuit breakers used to disconnect circuits to dispensing equipment?
   Answer: Reference:

9. Article 514 generally requires areas where flammable liquids having a flash point below 100°F (38°C) are handled be classified as Class I locations. What is the flash point of diesel fuel?
   Answer: Reference:

10. Which section of the *NEC®* covers the bonding requirements for hazardous (classified) locations?
    Answer: Reference:

Chapter 10

Bulk Storage Plants

DEFINITION

A bulk storage plant is defined as a portion of the property where bulk fuels are received and distributed. While at that location, they may also receive additives or be blended to achieve the desired mixture requested. These fuels may be received and distributed in any number of ways, such as pipelines, railcar tanks, tank ships, or tank trucks. For further information, you may check the latest Flammable and Combustible Liquids Code NFPA 30 (ANSI) (see Figures 10–1 and 10–2).

The degree of hazard or flammability of liquids is commonly measured in terms of the flash point. The flash point is the minimum temperature at which the liquid gives off vapors sufficient to form an ignitable mixture with air near the surface of the liquid when tested under established procedures. Flammable liquids having a flash point less than 100°F require the greatest protection against accidental ignition. Gasoline, for example, has a flash point of about 45°F and is a very hazardous material under all climatic conditions. Diesel fuel has a flash point of 100°F. The flash point of jet fuels range from 10°F to 145°F. See Fire Hazard Properties in Flammable Liquid, Gas, and Volatile Solids, NFPA 325M.

CLASS I LOCATIONS

Article 514 applies in areas or rooms where Class I liquids are handled and stored and where gasoline is dispensed in conjunction with the bulk storage operations. The boundaries of these areas or rooms are defined by the floors, walls, roofs, or other solid partitions without communicating openings. *NEC®* Table 515–2 may be used to classify bulk storage facilities provided these areas meet all of the requirements of NFPA 30 and Article 501. The authority having jurisdiction may specify the extent of these spaces.

Figure 10–1   Photo of bulk storage tanks. (Courtesy of Appleton Electric Company)

Figure 10–2 Photo of a bulk storage and loading facility. (Courtesy of Crouse-Hinds, Division of Cooper Industries)

> *NEC*® Article 515 does not cover necessary physical and corrosion protection needs.

It is important that designers understand that the *NEC*® Chapter 5 articles deal with the special considerations necessary to achieve a safe system in hazardous locations. However, other considerations may be necessary to achieve an installation that will last many years in the applicable environment. As an example, in many bulk storage tank farms corrosion can be an important element to consider. Petroleum products and corrosive chemicals may greatly reduce the life of the wiring method if not considered. Physical protection may also be an important factor in these areas. When both are present, the wiring method options are greatly limited. Coated galvanized rigid conduit may be advisable since this wiring method provides both good physical protection as well as excellent corrosion protection.

Adequately ventilated indoor areas containing pumps, bleeders, withdrawal fittings, meters, and similar equipment that are connected to pipelines handling flammable liquids under pressure are considered as Class I Division 2 locations where such spaces are within a 5-foot radius in any direction of an exterior surface of such equipment. A Class I Division 2 location also extends 25 feet horizontally from the surface of such equipment and upward to a height of 3 feet above grade (see Figure 10–3). Where such spaces are inadequately vented, they are considered as a Class I Division 1 location. Other provisions affecting ventilating requirements are discussed in the flammable liquids code, NFPA 30. Outdoor areas containing the same equipment itemized in the preceding paragraph are to be considered as Class I Division 2 locations within a 3-foot radius in all directions from the exterior surfaces of such equipment. The Class I Division 2 location also extends 10 feet horizontally from any surfaces of such equipment and extends upward to a height of 18 inches above grade. In *NEC*® Table 515–2 and Figure 10–3, such spaces are considered to be Class I Division 1 locations within 3 feet in all directions of the vent pipe or fill pipe opening. Class I Division 2 locations are within the space extending in all directions between a 3-foot and 5-foot radius from the vent pipe or fill pipe opening and horizontally within the radius of 10 feet from the vent or fill pipe opening to a height of 18 inches above grade. In all indoor areas where volatile, flammable liquids are transferred to individual containers or where fire enforced ventilation is not reliably maintained, the entire area involved must be classified as a Class I Division 1 location (*NEC*® Table 515–2[b]).

More limited hazardous locations are also defined for outdoor areas where volatile, flammable liquids are transferred to individual containers as well as for indoor areas which have reliably maintained fire-enforced ventilation. In outside locations, the space extending 3 feet in all directions from a vehicle tank when loading through an open dome, or from a vent when loading through a closed dome having atmospheric venting, is classified as a Class I Division 1 location. Between a 3-foot and 5-foot radius, it is a

Figure 10–3 Bulk storage installation guidelines. (Courtesy of Crouse-Hinds, Division of Cooper Industries)

Class I Division 2 location. Hazardous areas surrounding loading facilities and tank facilities are defined in *NEC*® Table 515–2. The space within 3 feet in all directions of a fixed connection used in bottom loading or unloading is a Class I Division 1 location. In loading through a closed dome having atmospheric venting or in loading through a closed dome having a vapor-recovery system, it is considered as a Class I Division 2 location. Where bottom loading is used, the Class I Division 2 location covers a 10-foot radius from the point of connection to a height of 18 inches above grade (see Figure 10–3). In aboveground tanks, the spaces between the shell and above the roof of floating roof tanks are Class I Division 1 locations. The area within 10 feet of where the shell sides ends and the roof of the aboveground tank that does not have a floating roof, is classified as a Class I Division 2 location. Where dikes are provided, the area within the dike is also a Class I Division 2 location. Any point within a 5-foot radius of a vent opening for an aboveground tank is a Class I Division 1 location. The space between a 5-foot and 10-foot radius from the vent of an aboveground tank is a Class I Division 2 location. Hazardous locations near fill pipes and vent pipes for underground tanks are defined in *NEC*® Table 514–2 (see Figure 10–3). Any pit or depression that lies wholly or partly within either a Class I Division 1 or Class I Division 2 location must be classified entirely as a Class I Division 1 location unless provided with forced draft ventilation. Where such forced-draft ventilation is provided, all such spaces may be classified as a Class I Division 2 location. Any pit or depression that contains gasoline piping, valves, or fittings is deemed a Class I Division 2 location when located in areas that would not otherwise be classified as a Class I area, such as where a gasoline pipeline or the pit outside the bulk storage plant contains meters or valves. Storage and repair garages for tank trucks are classified as commercial garages; that is, the Class I Division 2 area extends up to a height of 18 inches above grade unless the enforcement authority rules that the hazardous area must extend higher than 18 inches. Wiring within hazardous areas is required to comply with the *NEC®* requirements for Class I Division 1 or Class I Division 2.

> **Warning:** Great care must be exercised around waterways when loading and unloading flammable fluids. Careful consideration must be given to the hazards involved, plus corrosion and environmental concerns (see Figure 10–4).

CLASS I DIVISIONS 1 & 2 WIRING AND EQUIPMENT

All defined classified locations shall be in compliance with the provisions of *NEC®* Table 515–2.

Above the hazardous areas, fixed equipment installed in bulk storage plants is required to be of the totally enclosed type if capable of producing arcs or sparks, unless provided with screens or guards to prevent the escape of the hot particles produced. All portable equipment, such as portable lamps and their cords, is required to comply with Article 501. Fixed wiring shall be in metal raceways, Schedule 80 rigid

Figure 10–4 Marine wharf or loading and unloading dock area. (Reprinted with permission from NFPA 70–1996, *National Electrical Code®*, Copyright © 1995, National Fire Protection Association, Quincy, Massachusetts 02269. This reprinted material is not the complete and official position of the NFPA on the referenced subject, which is represented only by the standard in its entirety)

nonmetallic conduit, or in cable types MI, MC, TC, or SNM.

Underground wiring is required to be installed with threaded rigid metal conduit or threaded steel intermediate conduit, or where buried not less than 2 feet below the earth, is permitted to be in rigid nonmetallic conduit or an approved cable. Where rigid nonmetallic conduit is used, threaded rigid metal conduit or threaded intermediate metal conduit shall be used the last 2 feet of the conduit run to emergence to the point of connection to the aboveground raceway. Where cable is used, it shall be enclosed in threaded rigid metal conduit or threaded steel intermediate metal conduit from the point of the lowest buried cable level to the point of connection above the raceway.

> Conductor insulation in all underground wiring must comply with Section 501–13. An equipment-grounding conductor to ground all noncurrent-carrying metal parts is required in all nonmetallic wiring methods.

Sealing

Sealing fittings are required in electrical systems in accordance with the *NEC®* provisions for Class I areas, that is, in all raceways entering or leaving a hazardous location, all raceways entering or leaving an arcing device, and all raceways 2 inches in size or larger entering or leaving a junction box and containing terminals, taps, or splices.

Nonhazardous Locations

Office areas, boiler rooms and similar locations that are beyond the limits of the hazardous area are not required to come within the special wiring rules of this article.

GROUNDING

Article 250 applies to all areas within the scope of Article 515. Class I areas must also comply with Section 501–16.

Static electricity can build up to dangerous levels from truck tires and should be considered. Manufacturers can provide a safe means for static discharge.

REVIEW QUESTIONS

(Answers to some of these questions may be found in the introduction, previous chapters, and 1996 *NEC®*.)

1. Pits and depressions located in bulk storage plants without ventilation are classified _____ .
   Answer: Reference:

2. Is type NM cable an acceptable wiring method above the Class I areas in bulk storage facilities?
   Answer: Reference:

3. Are seals required in Schedule 80 rigid nonmetallic conduit runs?
   Answer: Reference:

4. Does Section 250–78 apply to all of Article 515?
   Answer: Reference:

5. Indoor equipment installed in accordance with NFPA 30 where flammable vapor air mixtures may exist under normal operations must be wired Class I Division 2 and _____.
   Answer: Reference:

6. Gasoline fuel-dispensing equipment must meet the requirements of _____.
   Answer: Reference:

7. Seals must be installed in accordance with Section 515–6 and _____.
   Answer: Reference:

8. The area adjacent to open sumps in piers or wharves is to be classified _____.
   Answer: Reference:

9. Drum and container filling areas are to be classified Class I Division 1 in an area _____ circumference of the vent and fill openings.
   Answer: Reference:

10. All buried raceways under the defined Class I locations must meet the requirements of a Class I Division _____ location.
    Answer: Reference:

Chapter 11

Spray Application, Dipping, and Coating Processes

DEFINED LOCATION

> These locations include both Class I and Class II areas depending on the application.

The application of flammable liquids, combustible liquids, and combustible powders by spray operations and the application of flammable liquids or combustible liquids at temperatures above their flash point by dipping, coating, and other means regularly or frequently are covered in *NEC*® Article 516. The classification is based on quantities of dangerous vapors, combustible dusts, mists, residues, or deposits.

> Additional information regarding the safeguards for these processes is to be found in NFPA 33–1996, NFPA 34–1996, and NFPA 91–1990.

These finishes may be applied by means of dipping, brushing and spraying, pneumatic spraying, fixed electrostatic spraying, electrostatic hand spraying, electrostatic fluidized beds, and powder spraying with electrostatic power spraying guns.

Two distinct hazards are present in paint spraying operations: explosive mixtures in the air from the spray and its vapor, and the combustible residue which may accumulate inside the booth (see Figures 11–1 and 11–2).

Class I or Class II Division 1 Locations

Section 516–2(a) states that the interior of spray booths and their exhaust ducts and all spaces within 20 feet horizontal in any direction from the spraying operation are classified as a Class I Division 1 location. The areas where spraying more extensive than touch-up spraying operations are performed and all space within 20 feet horizontally in any direction of dip tanks and their drainboards are classified as a Class I Division 1 location.

These particular facilities may contain both Class I and Class II requirements, depending upon the material which is being applied.

- Interiors of spray booths and rooms
- The interior of exhaust ducts
- Areas in the direct path of spray operations
- Areas for dipping and coating operations
- All space within 5-foot radial distance from the vapor sources extending from these surfaces to the floor

The vapor source shall be a liquid exposed in the process and the drainboard and any dipped or coated object from which it is possible to measure the vapor

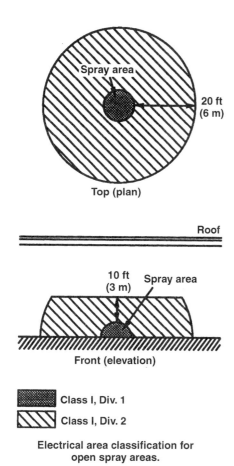

Figure 11–1 Class I or Class II location adjacent to an unenclosed spray operation. (Reprinted with permission from NFPA 70–1996, *National Electrical Code®*, Copyright © 1995, National Fire Protection Association, Quincy, Massachusetts 02269. This reprinted material is not the complete and official position of the NFPA on the referenced subject, which is represented only by the standard in its entirety)

for concentrations exceeding 25 percent of the lower flammable limit at a distance of 1 foot in any direction, pits within 25 feet horizontally of the vapor source (if the pit extends beyond 25 feet, a vapor stop must be provided or the entire pit must be classified), and the interior of any enclosed coating or dipping process.

Class I or Class II Division 2 Locations

These locations include the following spaces, again, depending upon the material being applied, as to whether it would be Class I or Class II location: open spraying, all space outside but within 20 feet horizontally and 10 feet vertically of the Division 1 location as

Figure 11–2 Class I or Class II Division 2 location adjacent to a close top, open face or open front spray booth or room. (Reprinted with permission from NFPA 70–1996, *National Electrical Code®*, Copyright © 1995, National Fire Protection Association, Quincy, Massachusetts 02269. This reprinted material is not the complete and official position of the NFPA on the referenced subject, which is represented only by the standard in its entirety)

defined and not separated by partitions (see Figure 11–1). For spraying operations conducted within an enclosed spray booth or room, the space shown in Figure 11–2 and the space within 3 feet in all directions from the openings other than the open face or front shall be considered a Class I and II Division 2. Class I and II Division 2 locations shown in Figure 11–2 shall extend from the spray booth or room as follows: If the ventilation system is interlocked with the spraying equipment so as to make the spraying equipment inoperative when the ventilation system is not in operation, the space shall extend 5 feet from the open face or front of the booth or room, as otherwise shown in Figure 11–2. When the ventilation system is not interlocked with the spraying equipment, the space shall extend 10 feet from the open face or front of the spray booth (see Figure 11–2). For spraying operations conducted within an open top spray booth, the space 3 feet vertically above the booth and within 3 feet of other booth openings shall be considered as a Division 2 location. For spraying operations confined to an enclosed booth or room, the space within 3 feet in all directions from any openings must be considered as a Division 2 location as shown in Figure 11–2. For dip tanks and drainboards, the space 3 feet above the floor and extending 20 feet horizontally in all directions from the Division 1 location shall be classified Class I or II Division 2, except that the space would not be required to be hazardous where the vapor source area is 5 square feet or less and where the contents of the open tank, trough, or container does not exceed 5 gallons. For the classification of open processes, see Figure 11–5. For the classification of open top containers, see Figure 11–6. In addition, the vapor concentration during operation shutdown periods shall not exceed 25 percent of the lower flammable limit outside the Class I location as specified in *NEC®* Section 516–2(a)(4). In an enclosed coating and dipping operation (see Figures 11–4 and 11–7), the space adjacent to these operations is not considered to be classified except within 3 feet in all directions of any opening or enclosure which must be classified as a Division 2 location (see Figure 11–3).

Nonhazardous Locations

Adjacent areas which are cut off by tight partitions, without communicating openings, and within which hazardous vapors or combustible powders are not likely to be released, may not be required to be classified. Areas utilizing drying, curing, or fusion apparatus provided with positive mechanical ventilation adequate to prevent accumulation of flammable concentrations of vapors and provided with effective interlocks to de-energize all electrical equipment other

Electrical area classification for enclosed spray booth or spray room.

Figure 11–3 Class I or Class II Division 2 location adjacent to openings in an enclosed spray booth or room. (Reprinted with permission from NFPA 70–1996, *National Electrical Code®*, Copyright © 1995, National Fire Protection Association, Quincy, Massachusetts 02269. This reprinted material is not the complete and official position of the NFPA on the referenced subject, which is represented only by the standard in its entirety)

than the equipment approved for Class I locations, may be classified as nonhazardous where the authority having jurisdiction so judges (see Figure 11–3).

Information pertaining to drying ovens or furnaces may be found in NFPA 86–1990.

Wiring and Equipment in Class I Locations

All electric wiring within the classified location as defined in Section 516–2 where *only vapors* are

Figure 11–4 The extent of a Class I Division 1 and Class I Division 2 location for an open-dipping process. (Reprinted with permission from NFPA 70–1996, *National Electrical Code®*, Copyright © 1995, National Fire Protection Association, Quincy, Massachusetts 02269. This reprinted material is not the complete and official position of the NFPA on the referenced subject, which is represented only by the standard in its entirety)

Figure 11–5 Area classification for open process operations. (Reprinted with permission from NFPA 70–1996, *National Electrical Code®*, Copyright © 1995, National Fire Protection Association, Quincy, Massachusetts 02269. This reprinted material is not the complete and official position of the NFPA on the referenced subject, which is represented only by the standard in its entirety)

contained *(no residue)* must comply with the applicable provisions of *NEC®* Article 501. There shall be no electrical equipment located where dangerous deposits of combustible or flammable vapors, mists, residue, dusts, or deposits may be present unless specifically "listed" for the location and purpose. Only electrical wiring in rigid metal conduit or intermediate metal conduit with metal boxes is permitted. The boxes are not permitted to contain terminations, splices, or taps.

Illumination

Specific requirements for lighting in these areas can be found in Section 516–3 and NFPA 33. Paint spray booth manufacturers have struggled to meet the needs while complying with the requirements of this *NEC®* section. New and innovative methods have been employed to provide safety while meeting the needs of the customer. It has generally been the accepted rule that general-purpose fixtures could be used if they were installed away from openings to the booth and nonbreakable clear panels isolated the fixture from the booth. Additionally no access to service the fixture was required from inside the booth, and there was assurance that there would be no heat buildup on the inside face of the panels. However, this required that the booth have workable clearances around the outside of

Spray Application, Dipping, and Coating Processes   109

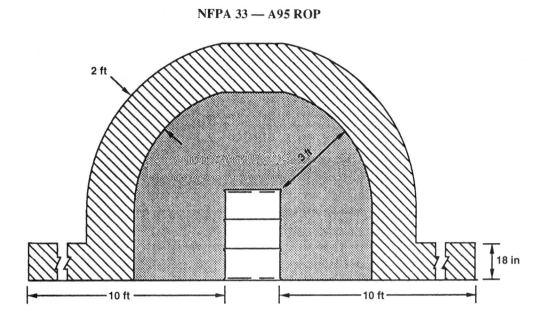

Figure 11–6 Area classification for open-top containers. (Reprinted with permission from NFPA 70–1996, *National Electrical Code*®, Copyright © 1995, National Fire Protection Association, Quincy, Massachusetts 02269. This reprinted material is not the complete and official position of the NFPA on the referenced subject, which is represented only by the standard in its entirety)

Figure 11–7 Area classification for enclosed processing operations. (Reprinted with permission from NFPA 70–1996, *National Electrical Code*®, Copyright © 1995, National Fire Protection Association, Quincy, Massachusetts 02269. This reprinted material is not the complete and official position of the NFPA on the referenced subject, which is represented only by the standard in its entirety)

Figure 11–8 Example of a light fixture, frame, and seal installation in a paint spray box.

the booth and that was not always obtainable. NFPA 33 now provides reduced clearances and clear directions on these installation guidelines. The authority having jurisdiction should always be consulted before building or installing a booth that varies from the general requirements (see Figures 11–8 and 11–9).

Portable Equipment

Generally portable equipment is not permitted to be used in a spray area during spraying operations.

- Section 516–3(d) Portable electric lamps or other utilization equipment shall not be used in a spray area during spray operations.
- Exception No. 1: Where portable electric lamps are required for operations in spaces not readily illuminated by fixed lighting within the spraying area, they shall be of the type approved for Class I Division 1 locations where readily ignitable residues may be present.
- Exception No. 2: Where portable electric drying apparatus is used in automobile refinishing spray booths and the following requirements are met: (1) the apparatus and its electrical connections are not located within the spray enclosure during spray operations; (2) electrical equipment within 18 inches (45.7 cm) of the floor is approved for Class I Division 2 locations; (3) all metallic parts of the drying apparatus are electrically bonded and grounded; and (4) interlocks are provided to prevent the operation of spray equipment while drying apparatus is within the spray enclosure to

Light fixture that is integral part of spray area and serviced from outside of spray area.

Figure 11–9 Sketch to show an acceptable method for installing illumination. (Reprinted with permission from NFPA 70–1996, *National Electrical Code*, Copyright © 1995, National Fire Protection Association, Quincy, Massachusetts 02269. This reprinted material is not the complete and official position of the NFPA on the referenced subject, which is represented only by the standard in its entirety)

- allow for a 3-minute purge of the enclosure before energizing the drying apparatus, and to shut off drying apparatus on failure of ventilation system.
- (e) Electrostatic Equipment. Electrostatic spraying or detearing equipment shall be installed and used only as provided in Section 516–4.
- (FPN): For further information, see Spray Application Using Flammable and Combustible Materials, NFPA 33–1989 (ANSI).

Special Equipment

The requirements for installing fixed electrostatic equipment, electrostatic hand spraying equipment, and the wiring of equipment for Class I or II locations and the areas above the Class I and Class II locations may be found in Section 516–4(a) through (j), 516–5(a) through (e), and 516–6(a) through (d). These requirements are extracted material from NFPA 33 and 34 and are very specific. These standards and the *NEC*® sections referenced must be followed precisely.

Wiring and Equipment above Class I and Class II Locations

All fixed wiring must be in metal raceways, rigid nonmetallic conduits, electrical nonmetallic tubing, or in types MI, MC, TC, or SNM cable where the wiring method leads into or through the classified Class I or Class II location. Suitable seals must be employed. Section 516–7(a) provides specific requirements. No equipment that may produce a spark or arc, such as lamps, lampholders, switches, motors, or other similar equipment, shall be located above classified areas where fresh goods are handled unless of the totally enclosed type. Specific requirements for these installations are found in Section 516–7(b).

Grounding

All metal raceways, metal jacketed cables, and all noncurrent-carrying metal parts of fixed or portable equipment must be grounded, regardless of voltage, and comply with the provisions of Article 250.

REVIEW QUESTIONS

(Answers to some of these questions may be found in the introduction, previous chapters, and 1996 *NEC*®.)

1. Is it possible for listed or approved equipment to meet the requirements of both Class I and Class II?
   Answer: Reference:

2. What does the word "listed" mean?
   Answer: Reference:

3. Standard light switches may not be located within _____ feet of openings to closed top spray booths.
   Answer: Reference:

4. Is it permissible to use EMT (Electrical Metallic Tubing) in locations above Class I and Class II locations?
   Answer: Reference:

5. Areas within 3 feet of openings in coating and dipping operations are classified as _____.
   Answer: Reference:

6. The interior of enclosed coating and dipping processes are classified as Class _____ or _____ Division _____.
   Answer: Reference:

7. Electrostatic spraying and detearing equipment must be installed in accordance with _____.
   Answer: Reference:

8. Objects being painted by electrostatic hand-spraying process must be in _____ contact with the conveyor or grounded support.
   Answer: Reference:

9. What is "nonincendive equipment"?
   Answer: Reference:

10. The degree of hazard associated with powder spraying or coating is dependent upon, the _____, _____, _____, _____, _____, _____, _____, _____, _____.
    Answer: Reference:

Chapter 12

Health Care Facilities, Residential and Custodial Facilities, and Limited Care Facilities

The requirements for these areas are found in *NEC®* Article 517. Additional information regarding the installation, testing, and safeguards of these areas may be found in NFPA 99–1990. Specific definitions of areas within these facilities may be found in *NEC®* Section 517–3.

Health care facilities include the electrical equipment installations in hospitals, nursing homes, extended care facilities, clinics, and dental offices. In addition, *NEC®* Article 517 includes provisions for electrical installations in areas provided for patients undergoing a type of treatment involving the use of electrical energized equipment connected to the heart. Requirements for essential and emergency electrical systems in hospitals have also been covered in this *NEC®* article. These specialized systems must meet exacting and detailed requirements, a discussion of which is beyond the scope of this manual.

The following comments are limited to the electrical requirements in special hazardous areas in hospitals, usually identified as an anesthetizing location, including hospital operating rooms, delivery rooms, anesthesia rooms, and any corridor, utility room, or other area which may be used for this purpose, or where flammable anesthetics are stored. The hazardous space and flammable anesthetizing locations extend from the floor to a height of 5 feet and are classified as Class I Division 1 locations. The entire room where flammable anesthetics are stored is classified as a Class I Division 1 location. All fixed wiring and equipment, and all portable equipment, including lamps and other utilization equipment, operating at more than 10 volts between conductors shall comply with the requirements of Sections 501–1 through 501–15 and Sections 501–16(a) and (b) for Class I Division 1 locations. All such equipment shall be specifically approved for the hazardous atmospheres involved.

Receptacles and attachment plugs in hazardous (classified) location(s) shall be listed for use in Class I Group C hazardous (classified) locations and shall have provision for the connection of a grounding conductor. Receptacles and attachment plugs located above hazardous anesthetizing locations shall be listed for hospital use for services of prescribed voltage, frequency, rating, and number of conductors with provision for the connection of the grounding conductor. This requirement shall apply to attachment plugs and receptacles of the 2-pole, 3-wire grounding type for single-phase 120-volt, nominal, AC service.

Each power circuit within, or partially within, a flammable anesthetizing location as referred to in Section 517–60 shall be isolated from any distribution system supplying other than anesthetizing locations. Maximum potential on the primaries or secondaries of isolating transformers is limited to 300 volts. There are some specific allowances in Section 517–61 that may apply.

Branch circuits which supply an anesthetizing location may not supply any other location. Special insulation on circuit conductors is also required to minimize capacitance of circuitry. These circuits are required to be supplied by the ungrounded side of an isolated transformer. Branch circuits supplying only fixed lighting fixtures in nonhazardous areas other than surgical lighting fixtures or approved permanently installed X-ray equipment may be supplied by grounded circuits. If supplied by a grounded circuit, the conductors must be in a separate raceway from the conductors of the ungrounded circuit. The lighting fixtures and X-ray equipment must be located 8 feet above the floor, with switches controlling the grounded circuits located outside the anesthetizing location.

Low-voltage current circuits are required by *NEC®* for electrical apparatus and equipment having exposed current-carrying parts that are frequently in contact with bodies of patients. Such equipment circuits must be designed and approved for operation of 8 volts or less. More detailed and complete information may be found in NFPA 99. **Note:** The NFPA 99 handbook will provide great assistance when designing or installing a health care facility.

REVIEW QUESTIONS

(Answers to some of these questions may be found in the introduction, previous chapters, and 1996 *NEC®*.)

1. Name five examples of flammable anesthetics.
   Answer: Reference:

2. In areas where flammable anesthetics are used, the entire area up to _____ feet above the floor is considered to be a Class I Division 1 location.
   Answer: Reference:

3. Areas where flammable anesthetics or flammable disinfecting agents are stored shall be considered to be a Class I Division 1 location from _____ to _____.
   Answer: Reference:

4. Anesthetic locations designated as wet locations must meet the requirements of Section _____.
   Answer: Reference:

5. Areas designated for the exclusive use of inhalation nonflammable anesthetics shall be designated as _____ _____ _____ _____ _____.
   Answer: Reference:

6. Surgical lighting located above hazardous anesthetic locations must meet the requirements of _____.
   Answer: Reference:

7. Equipment operating at not more than _____ volts between conductors need not be grounded.
   Answer: Reference:

8. Generally, circuits in flammable anesthetizing locations shall be _____ from a distribution system supplying other than anesthetizing locations.
   Answer: Reference:

9. Defined hazardous locations within health care facilities shall comply with the provisions of Articles _____ and _____.
   Answer: Reference:

10. Flexible cords used in hazardous areas for the connection of utilization equipment operating at more than _____ volts between conductors shall be a type approved for _____ _____ _____.
    Answer: Reference:

Chapter 13

Article 505 Class I Zone 0, 1, and 2 Locations

> **Notice**
> There is an appeal requesting that Article 505 and all related references for the alternative hazardous classification scheme be deleted from the 1996 *NEC®* (see page 20 1996 *NEC®*).

A new Article 505 has been added to the 1996 *NEC®* introducing the European zone concept as a new optional method for qualified registered professional engineers (see *NEC®* Section 500–6) for what have been the accepted Class I areas. It provides a new level of the classification of Class I hazardous (classified) locations. This alternate method uses the zone concept to meet the requirements for electrical equipment and wiring for all voltages in locations where fire or explosive hazards may exist due to flammable gases or vapors and flammable liquids.

The new article modifies the general rules of the code as it applies to the electric wiring and equipment in locations classified as Class I Zone 0, Zone 1, and Zone 2.

ZONES

As the zones are defined below, you will note that except for Zone 0 the classifications are similar to the respective divisions in Article 501. However, Zone 0 is the area within the previously classified Division 1 location. It is an area that traditionally has not contained any electrical wiring or equipment, such as inside vessels, even though it was allowed. Intrinsically safe systems offer a safe low energy method in these locations. However, intrinsically safe systems must be routed to unclassified areas; in many instances this may be difficult to achieve. In these instances where it is necessary, the traditional Class I Division 1 equipment and wiring methods can be used.

A Class I Zone 0 location is a location (1) in which ignitable concentrations of flammable gases or vapors are present continuously or (2) in which ignitable concentrations of flammable gases or vapors are present for long periods of time.

A Class I Zone 1 location is a location (1) in which ignitable concentrations of flammable gases or vapors are likely to exist under normal operating conditions, (2) in which ignitable concentrations of flammable gases or vapors may exist frequently because of repair or maintenance operations or because of leakage, (3) in which equipment is operated or processes are carried on of such a nature that equipment breakdown or faulty operations could result in the release of ignitable concentrations of flammable gases or vapors and also cause simultaneous failure of electrical equipment in a mode to cause the electrical equipment to become a source of ignition, or (4) that is adjacent to a Class I Zone 0 location from which ignitable concentrations of vapors could be communicated, unless communication is prevented by adequate positive-pressure ventilation from a source of clean air and effective safeguards against ventilation failure are provided.

A Class I Zone 2 location is a location (1) in which ignitable concentrations of flammable gases or vapors are not likely to occur in normal operation and if they do occur will exist only for a short period, (2) in which volatile flammable liquids, flammable gases, or flammable vapors are handled, processed, or used, but in which the liquids, gases, or vapors normally are contained within closed containers or closed systems from which they can escape only as a result of accidental rupture or breakdown of the containers or system, or as the result of the abnormal operation of the equipment with which the liquids or gases are handled, processed, or used, (3) in which ignitable concentrations of flammable gases or vapors normally are prevented by positive mechanical ventilation, but which may become hazardous as the result of failure or abnormal operation of the ventilation equipment, (4) that is adjacent to a Class I Zone 1 location from which ignitable concentrations of flammable gases or vapors could be communicated, unless such communication is prevented by adequate positive-pressure ventilation from a source of clean air and effective safeguards against ventilation failure are provided.

GROUPS

The groupings are also designated differently than the Class I groupings; however, they are approximately equivalent.

Group I electric apparatus is intended for use in mines susceptible to fire and dampness and is not part of this code; *NEC*®. Section 90–2(b)(2) specifically exempts underground mines from the provisions of the *NEC*®.

Group II is subdivided according to the nature of the liquid, gas, or vapor atmosphere for which it is intended.

- Group II A. Atmospheres containing acetone, ammonia, ethyl alcohol, gasoline, methane, propane, or gases or vapors of equivalent hazard. This group is similar to Class I Group D.
- Group II B. Atmospheres containing acetaldehyde, ethylene, town gas (coal gas) or gases or vapors of equivalent hazard. This group is similar to Class I Group C.
- Group II C. Atmospheres containing acetylene, hydrogen, carbon disulfide, or gases or vapors of equivalent hazard. This group is similar to Class I Groups A and B.

(FPN No. 1): This grouping in (c) above is approximately equivalent to Class I Group D as described in Section 500–3(a)(4).

(FPN No. 2): The gas subdivision in (a), (b), and (c) above is based on the maximum experimental safe gap, minimum igniting current or both. The test apparatus for determining the maximum experimental safe gap is described in ICE Publication 79–IA (1975) and UL Technical Report No. 58 (1993).

(FPN No. 4): It is necessary that the meaning of the different equipment markings and Group II classifications be carefully observed to avoid confusion with Class I Divisions 1 and 2, Groups A, B, C, and D.

EQUIPMENT

All equipment that has been approved for a Zone 0 location is permitted in a Zone 1 or Zone 2 location of the same gas group. Equipment that has been approved for a Zone 2 location shall be permitted in a Zone 2 location of the same gas group.

Equipment approved for use in this application is required to be marked to show the class, zone, gas group, and temperature class reference to a 40°C (104°F) ambient. The temperature class marked on equipment shall be as shown in Table 505–10(b).

NEC® Table 505–10(b).

Classification of Maximum Surface Temperature for Group II Electric Apparatus

Temperature Class	T1	T2	T3	T4	T5	T6
Maximum Surface Temperature (–C)	450	300	200	135	100	85

WIRING METHODS

In Class I Zone 0 locations, only the following wiring methods shall be used:

(1) Intrinsically safe wiring in accordance with Article 504.
(2) Threaded rigid metal conduit complying with the last paragraph of Section 500–2 with listed explosion-proof fittings and boxes or listed type MI cable with listed fitting suitable for use in Class I Division 1 locations. The conduit or cable shall contain only intrinsically safe or nonincendive circuits. All conduits ½-inch trade size and larger shall be sealed in accordance with Sections 501–5(a), (c), and (d).
   (FPN): In applying the sealing requirements of Sections 501–5(a), (c), and (d), it is intended that the references to Division 1 be interpreted as Zone 0.
(3) Nonconductive optical fiber cable with approved energy-limited supply.

In Class I Zone 1 locations, all wiring methods permitted for Class I Division 1 locations and Class I Zone 0 locations, including requirements for sealing, are permitted. Cables listed for the location are also permitted.

In Class I Zone 2 locations, all wiring methods permitted for Class I Division 2, Class I Division 1 or Division 2 and Class I Zone 0 or Zone 1 locations, excluding requirements for sealing, shall be permitted. Cables approved for hazardous (classified) locations are also permitted.

EQUIPMENT

In Class I Zone 0 locations, only equipment specifically listed and marked as suitable for the location shall be permitted.

In Class I Zone 1 locations, only equipment specifically listed and marked as suitable for the location shall be permitted.

Exception: Equipment approved for use in Class I Division 1 or Class I Zone 0 locations of the same gas group and with similar temperature marking, if any, shall be permitted.

In Class I Zone 2 locations, only equipment specifically listed and marked as suitable for the location shall be permitted.

Equipment suitable for use in Class I Division 1 or Division 2 or Class I Zone 0 or Zone 1 locations of the same gas group and with similar temperature markings, if any, shall be permitted.

GROUNDING AND BONDING

Grounding and bonding shall comply with Article 250 and Section 501–16.

SUMMARY

This is intended to be optional and in no way does it replace Article 501 classifications and methods. This method is introduced as an optional method which can be used under strict supervision by a qualified profession-registered engineer. It introduces to the *National Electrical Code®* and those who adopt the *NEC®* a concept that has been used internationally. For the impact and comparison, see Figures 13–1 and 13–2.

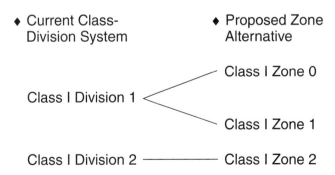

Figure 13–1 Proponents of this new article presented this comparison chart to explain the impact of adding Article 505 to the 1996 *NEC®*.

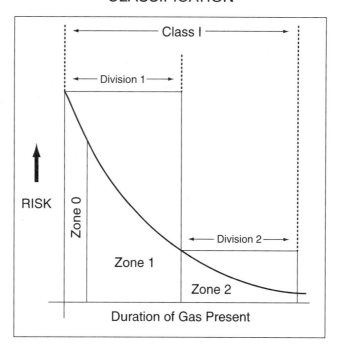

Figure 13–2 This is an example of how an area may be used as an alternative to classifying an area under the zone concept Article 505.

REVIEW QUESTIONS

(Answers to some of these questions may be found in the introduction, previous chapters, and 1996 *NEC®*.)

1. In a storage and repair facility for motorcycles where 10 or more machines will be repaired or worked on at the same time, with compliance with *NEC®* Section 511–2 be mandated for the purpose of classifying the facility.
   Answer: True or False.   Reference:

2. Does *NEC®* Section 514–6(a) require that the first fitting be a seal-off device where the conduit emerges from the ground/concrete at underground pits where gasoline tanks are located and have submersible pumps?
   Answer: True or False.   Reference:

3. A business which specializes in oil changes for gasoline-powered vehicles has inspection pits and service bay areas wherein personnel can have access to the under portions of the vehicles. Extension cords, portable lights, and portable power tools are used in these service bays and also in the pits. Are the pits classified? If so, what classification would these pits have, and under what conditions?
   Answer: Reference:

4. I would like to install paddle fans in an aircraft hangar. The installation will be above 5 feet and within 10 feet of the wings and power plants of the aircraft housed within. Is this permissible? If not, where could these paddle fans be located without violating the requirements of the *National Electrical Code®*?
   Answer: Reference:

5. Exposed metal raceway surfaces in general patient care areas in a health care facility are not permitted to exceed a _____ differences at frequencies of 1000 Hertz or less measured across a 1000 ohm resistance.
   a. 100
   b. 250
   c. 500
   d. 1000
   Answer: Reference:

6. An underground gasoline tank remote pump from the dispenser is listed for a Class I Division 1 location, with the pump having a factory seal provided between the terminal housing and the motor. Is a listed seal-off fitting required by Section 501–5(a)(1) for this underground gasoline pump?
   Answer: Reference:

7. A commercial garage has a car showroom attached to it. There is only one doorway between them and no fire separation. May the showroom be wired in accordance with the first three chapters of the Code, such as in EMT nonmetallic sheath cable or type AC cable? What if the garage portion of the building has four air changes per hour?
   Answer: Reference:

8. A diesel dispenser is mounted adjacent to a gasoline dispenser on a service station island. A sealing fitting is installed in the conduit entering the gasoline dispenser. Is a sealing fitting required in the conduit entering the diesel fuel dispenser?
   Answer: Reference:

9. A rigid metal conduit runs under a commercial garage floor and extends up through the 18-inch hazardous area with no couplings or fittings in that 18-inch area. Is a seal-off fitting required in this conduit?
   Answer: Reference:

10. If Article 505 is used, area, classification, wiring, and equipment selection shall be under the supervision of a _____ _____ _____ _____.
    Answer: Reference:

Chapter 14

Intrinsically Safe Systems

WHAT IS INTRINSIC SAFETY?

Intrinsic safety is an explosion-prevention design technique applied to electrical equipment and the wiring of hazardous locations where flammable or combustible material is present. The technique is based upon limiting electrical and thermal energy levels below that which is required to ignite the specific hazardous atmospheric mixture present. Intrinsically safe wiring shall not be capable of releasing sufficient electrical or thermal energy under normal or abnormal conditions to cause ignition to a specific flammable or combustible atmospheric mixture in its most easily ignitable concentration. It is important to define the class and group for which any proposed intrinsically safe electrical circuits are to be installed. Since intrinsic safety is a technique for the worst case hazardous locations, consideration of the division is not necessary. Because intrinsic safety maintains energy levels below that required to ignite specific hazardous mixtures, it is important to know what the energy allowances are for operational and safety considerations. Intrinsically safe systems, Article 504, was introduced as an acceptable wiring system in the 1990 *National Electrical Code®*. Careful interpretation and application of this article will provide safe installations that are free from hazard.

DEFINITION

Article 504 covers the installation of intrinsically safe apparatus, wiring, and systems for Class I, II, and III locations. The specific and unique definitions of terms for this article appear in Section 504–2.

The use of intrinsically safe equipment is primarily limited to process control instrumentation. Since these electrical systems lend themselves to the low energy requirements, ANSI/UL 913–1988 provides information on the design test conditions and evaluation of safe apparatus and associated apparatus for use in Class I, Class II, and Class III Division 1 hazardous classified locations. Underwriters Laboratories Inc. and Factory Mutual list several devices in this category.

The equipment and associated wiring must be installed so they are positively separated from the non-intrinsically safe circuits. It is very important that these systems be designed and installed precisely as per the control drawings. It is also very important that all equipment be marked with the control drawing number. Following these guidelines will help to prevent corruption of the integrity of the system. Induced voltages could defeat the concept of intrinsically safe systems.

NEC® Section 504–1 clarifies that this article is only intended to cover intrinsically safe apparatus and wiring in Class I, II, and III locations and not intrinsically safe systems that are not located in a classified area. All intrinsically safe apparatus and associated apparatus must be approved and accept as modified in Article 504. All applicable articles of the Code are applied to these systems. It is important to note that in applying *NEC®* Article 504, only the entire intrinsically safe system must be installed in accordance with control drawings, as required in *NEC®* Section 504–10. The control drawing identification must be marked on all equipment and apparatus. Generally, intrinsically safe systems may be permitted using any wiring method suitable for unclassified locations. However, sealing shall be provided in accordance with *NEC®* Section 504–70 and separation of intrinsically safe systems shall be provided in accordance with *NEC®* Section 504–30.

WIRING METHODS

General wiring methods are permitted. This would include any of the wiring methods in Chapter 3 that are appropriate for the location and use.

> **Warning:** Corrosion and physical protection have not been considered, only circuit safety. It is the designer's responsibility to consider all facets of the installation for performance and safety.

It should be noted that intrinsically safe systems are designed to always fail in the safe position. Should a wire or cable be crushed or cut accidentally, the operation of the equipment must fail safe in every instance. No arcing or sparking will occur due to the design and nature of intrinsically safe systems. Designers generally require physical protection, such as metal raceways, to protect intrinsically safe conductors and to provide assurance from costly accidental shutdowns and operational stoppages.

SEALING

NEC® Section 504–70 requires that conduits and cables be sealed in accordance with the hazardous location in which they are installed, such as Article 501 for Class I hazardous flammable vapors and gases. Sealing of intrinsically safe systems must be provided as specified in Section 501–5. For flammable or combustible dust, sealing is required as specified in *NEC®* Section 502–5. Seals are not required for enclosures that contain only intrinsically safe apparatus.

IDENTIFICATION

Permanent identification "Intrinsic safety wiring" is required at all terminals and throughout the entire length of the wiring method and support system at intervals no more than 25 feet apart. This requirement applies within the hazardous area and in the nonhazardous locations. Industry practice has been to identify this system with the color light blue for the cable systems; however, color coding is not required. Color coding is permitted where no other conductors used are light blue in color. For specific requirements, see *NEC®* Section 504–80(a) and (b). This strict marking identification is necessary in both hazardous and nonhazardous locations to prevent inadvertently mixing other systems with intrinsically safe systems, both during the initial installation and at a later date.

ELECTRICAL AND MECHANICAL SEPARATION

Open wiring of conductors and cables is permitted but must be separated from nonintrinsically safe circuits at least 50 mm and secured. It must also be secured and separated a minimum of 2 inches from conductors or cables of other systems not intrinsically safe. Intrinsically safe conductors generally shall not be placed in the same raceway, cable tray, or cable with circuits of other nonintrinsically safe systems. Exceptions to this general rule can be found in *NEC®* Section 504–30(a)(1) and (2).

Within enclosures, conductors must be secured and separated a minimum of 2 inches from conductors or cables of other systems not intrinsically safe. Separate wiring compartments or physical barriers are preferred by most designers.

Intrinsically safe circuits shall be separated from other intrinsically safe circuits by being placed in separate cables or with a grounded metal shield or by a minimum of 0.01 inches of insulation.

GROUNDING

All intrinsically safe systems associated apparatus, enclosures, raceways, cables, and cable shields must be grounded. The control drawings may also require supplementary bonding to the grounding electrode to some associated apparatus such as intrinsic safety barriers.

SAFETY ASSURED

There are a number of ways to design systems to assure intrinsic safety. However, all provide separation by an electrical barrier (associated apparatus) which separates intrinsically safe protected wiring from the unprotected nonintrinsically safe wiring. In all cases these barriers will limit energy into the hazardous location under unique specified fault conditions. The barriers generally fall into two basic categories: isolated and nonisolated.

Intrinsic safety barriers are passive networks consisting of voltage-limiting diodes and current-limiting resistors or semi-conductive current-limiting components. They allow the electrical signal to flow while limiting the energy level. These are usually protected from opening a high fault condition by a fuse. To provide over-voltage protection, intrinsic safety barriers must be grounded in accordance with the control drawings and must be located outside the classified hazardous area.

Intrinsic safety galvanic isolation devices provide isolation between intrinsic safe wiring and nonintrinsic safe wiring without the necessity of a ground connection for over-voltage. These are generally equipped with signaling condition devices. Although this new accepted installation system will provide safe installations, it is paramount that they be installed precisely as designed and shown on the control drawings as permitted in *NEC®* Article 504. Manufacturers of these systems should provide additional installation and startup guidance to the installer.

REVIEW QUESTIONS

(Answers to some of these questions may be found in the introduction, previous chapters, and 1996 *NEC®*.)

1. General purpose equipment can be used with intrinsically safe systems except it must be labeled with the _____.
   Answer: Reference:

2. Is color coding of conductors required?
   Answer: Reference:

3. If conductors are color coded, they must be _____ and no other conductors are permitted to be _____.
   Answer: Reference:

4. Does Section 250–78 apply to hazardous areas where intrinsically safe systems are used?
   Answer: Reference:

5. Intrinsically safe systems in Class I Division 2 locations must be sealed in accordance with _____.
   Answer: Reference:

6. Section 504–30(b) contains specific requirements for "different intrinsically safe circuit conductors." What are "different intrinsically safe circuits"?
   Answer: Reference:

7. Intrinsically safe conductors in open wiring must be separated at least _____ inches from non-intrinsically safe conductors or enclosed in approved raceways or cables.
   Answer: Reference:

8. Is it permissible to ground an intrinsically safe system to a separate grounding electrode system and isolate it from the grounding electrode system used for the building service?
   Answer: Reference:

9. Intrinsic safety barriers are referred to as _____ _____.
   Answer: Reference:

10. All associated apparatus and intrinsically safe apparatus must be approved. What does "approved" mean?
    Answer: Reference:

Chapter 15

Wiring Methods for Hazardous Locations

Before selecting a wiring method, many factors must be considered: environment, physical protection requirements, temperature, and all other pertinent conditions which may affect the material selected. Once this has been done, a wiring method may be selected from Chapter 3 of the *NEC*®. Each installation generally requires several wiring methods to meet all of the conditions, such as a method that is acceptable for "dry" concealed locations, or a method that provides "physical protection." It may also be necessary to select a "flexible" wiring method for equipment connection, and a material that can be used in "wet locations" and "underground." There are wiring methods which are acceptable in all locations. The final choice must comply with the requirements of *NEC*® Sections 110–3(a) and (b) and is generally made by the designer. Once the wiring method or methods have been selected, it is the responsibility of the installer to comply with the provisions in *NEC*® Article 300 and the *NEC*® article that specifically covers each wiring method chosen before making the installation (see Figure 15–1).

> **Note:** General wiring methods are all found in the *NEC*®, Chapter 3. However, when wiring a hazardous location, the provisions of *NEC*® Chapter 5 generally limit the number of different types of wiring methods.

PROTECTION AGAINST PHYSICAL DAMAGE

The function of an electrical raceway is to facilitate the insertion and extraction of the conductors, and to protect them from mechanical injury. All raceway types facilitate easy insertion and extraction of conductors. However, only rigid metal conduit, intermediate metal conduit, and rigid nonmetallic conduit Schedule 80 provide protection of conductors against physical damage (see Figure 15–2). Electrical metallic tubing provides physical protection, except where severe physical damage to the conductors is likely.

The designer and installer can no longer rely on all raceways to provide physical protection. Steel raceways, rigid metal conduit as covered in *NEC*® Article 346 and intermediate metal conduit as covered in *NEC*® Article 345 provide the greatest physical protection and are permitted in all applications of the code (see Figures 15–3 and 15–4). Schedule 80 PVC as covered in *NEC*® Article 347 does provide excellent physical protection. However, a fine print note in *NEC*® Section 347–2 warns us that in cold temperatures PVC may become brittle and crack or break under impact. Extreme cold may cause some nonmetallic conduits to become brittle and therefore more susceptible to damage from physical contact. Therefore, Schedule 80 may be used for physical protection, but consideration must be given to the environment in which it is to be

Figure 15–1 Conduit nipples, elbows, and fittings. (Courtesy of Picoma Industries Inc.)

installed as well as the physical protection needed under those conditions. Rigid nonmetallic conduit is only permitted to be used in hazardous locations in very limited and specific locations. Electrical metallic tubing as covered in *NEC* Article 348 provides great physical protection for all but severe physical damage locations. EMT, like nonmetallic conduit, is only permitted on a limited basis in hazardous locations.

Raceways are known to frequently outlast repeated reconstruction and remodeling of the buildings they serve, and all these things must be considered when the designer is selecting the raceway for the specific installation. Great care is needed in making these decisions because one of the important contributing factors in selecting a raceway is the long-term economy of the installation. In the case of electrical modernization, old wires can often be pulled out and new wires pulled in without structural remodeling. It is up to the designer to select a raceway type that will provide this feature. *NEC* Section 300–4 requires conductors to be adequately protected where subject to physical damage in both exposed and concealed locations, except for intermediate metal conduit, rigid metal

Wiring Methods for Hazardous Locations 129

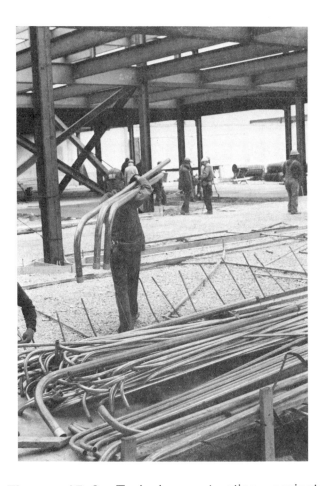

Figure 15-2 Typical construction project installing steel raceways which provides excellent physical protection which is needed in a concrete slab during the installation pour. (Courtesy of Allied Tube & Conduit)

Figure 15-3 A typical construction project installing conduit in a concrete slab; this is an excellent example of how physical damage could occur during a concrete pour. (Courtesy of Allied Tube & Conduit)

Figure 15-4 Test result summary. The after-impact internal diameter exceeded the 40% allowable fill in all tests for both rigid steel and IMC conduit. Conductors were not damaged in either rigid steel or IMC and in all cases conductors were easily extracted. (Courtesy of Allied Tube & Conduit)

conduit, rigid nonmetallic conduit, and electrical metallic tubing. Raceways and cables shall be protected where run through bored holes or notches in wood, through metal framing members, and where run parallel to the framing members where a distance of 1¼ inches from the edge of the framing member cannot be maintained. The raceway or cable type shall be protected from penetration by screws or nails with a steel plate sleeve or equivalent of at least 1/16-inch thick. It is the responsibility of the designer to select a raceway type that will provide protection to the conductor in the environment for which it is installed. The degree of physical damage that the wiring method may be subjected to should dictate the raceway type to be selected. For installations over 600 volts nominal, aboveground installations, the conductors shall be protected by installing them in rigid metal conduit, intermediate metal conduit, or rigid nonmetallic conduit. Conductors may also be run in cable trays, bus ways, or cable bus in some installations. The installation requirements for over 600 volt installations are found in *NEC*® Article 710. Section 710–3(a) governs aboveground installations, and Section 710–3(b)(1) provides the physical protection requirements for underground installations over 600 volts where they emerge from the ground.

Although raceways were originally designed to protect conductors from mechanical injury, as well as to provide the versatility for future changes by facilitating the insertion and extraction of conductors, in the past 30 years specialty raceways have been developed, accepted, and included in the *NEC®*.

UNDERGROUND REQUIREMENTS

NEC® Section 300–5 covers the requirements for underground installations. *NEC®* Section 300–5 gives the minimum cover requirements for cable or conduit to be installed (see Figure 15–5). All underground installations shall be grounded and bonded in accordance with *NEC®* Article 250. Where approved for direct burial, cable installed under a building shall be in a raceway that must extend beyond the outside wall of the building. Direct buried cables and cables emerging from the ground shall be protected from physical protection in a raceway extending from the minimum cover distance required in *NEC®* Section 300–5 below grade to a point at least 8 feet above finish grade. Conductors entering a building shall be protected to the point of entrance. Where a raceway or enclosure is subject to physical damage, the conductors must be installed in rigid metal conduit, intermediate metal conduit, or a Schedule 80 rigid nonmetallic conduit or equivalent. Raceways or cables shall not be backfilled with a material that will damage, prevent adequate compacting, or contribute to corrosion of raceways or cables. When it is necessary to prevent physical damage from occurring to a raceway or cable, the raceway or cable may be covered with select material (sand), suitable running boards, sleeves, or other approved means. Raceways through which moisture may contact energized live parts, such as conductors entering the building from outside into panelboards or switchboards which may contact live parts, shall be sealed or plugged at either or both ends. If the presence of hazardous gases or vapors are present, this may also require the sealing of underground raceways where they enter the building. A bushing or terminal fitting shall be placed on the end of a conduit or raceway which terminates underground where cables approved for direct burial, such as type "UF," emerge as a direct-burial wiring method. A seal incorporating the physical protection characteristics of a bushing is permitted in lieu of the bushing. *NEC®* Article 710 has similar requirements for underground installations of over 600 volts nominal (see Figure 15–6). The provisions of Section 710–3(b) and Table 710–3(b) provide specific requirements for those installations.

EXPOSURE TO DIFFERENT TEMPERATURES

Raceways exposed to different temperatures are covered in *NEC®* Section 300–7. This section of the *NEC®* requires that where portions of interior raceways are exposed to widely different temperatures, such as in refrigeration or cold storage plants, circulation of air from the warmer to the colder section through a raceway shall be prevented by sealing the open ends of the wiring method where the temperature change occurs. This is to reduce condensation from building up in the equipment being served by that raceway. Although the *NEC®* only requires this precautionary measure for interior installations, it is also recommended where wiring methods pass from outside to environmentally controlled areas inside a building (see Figure 15–7). Good installation practices will dictate entry into equipment in a manner so that the condensate will not drip on exposed live parts. It is not uncommon for improper installations to short out causing expensive downtime when this occurs. Special fittings designed for sealing conduit openings are not required for this application. Field methods for sealing may be accomplished using "duct-seal" or other removable putty which will not deteriorate conductor insulation. Expansion joints are required where necessary to compensate for thermal expansion and contraction. The specific requirements for nonmetallic raceways, which have a higher coefficient of expansion, are covered below (see Figure 15–8).

EXPANSION PROPERTIES OF RACEWAYS

Like all construction materials, raceway systems will expand and contract with variations in temperatures. The coefficient of linear expansion for various conduits is:

a. PVC conduit 3.38×10^5 in./in./°F,
b. Aluminum conduit 1.2×10^5 in./in./°F,
c. Steel conduit 0.6×10^5 in./in./°F.

Wiring Methods for Hazardous Locations 131

Table 300-5. Minimum Cover Requirements, 0 to 600 Volts, Nominal, Burial in Inches
(Cover is defined as the shortest distance measured between a point on the top surface of any direct buried conductor, cable, conduit, or other raceway and the top surface of finished grade, concrete, or similar cover.)

Location of Wiring Method or Circuit	Type of Wiring Method or Circuit				
	1 Direct Burial Cables or Conductors	2 Rigid Metal Conduit or Intermediate Metal Conduit	3 Nonmetallic Raceways Listed for Direct Burial without Concrete Encasement or Other Approved Raceways	4 Residential Branch Circuits Rated 120 Volts or Less with GFCI Protection and Maximum Overcurrent Protection of 20 Amperes	5 Circuits for Control of Irrigation and Landscape Lighting Limited to Not More than 30 Volts and Installed with Type UF or in Other Identified Cable or Raceway
All Locations Not Specified Below	24	6	18	12	6
In Trench Below 2-Inches Thick Concrete or Equivalent	18	6	12	6	6
Under a Building	0 (In Raceway Only)	0	0	0 (In Raceway Only)	0 (In Raceway Only)
Under Minimum of 4-Inch Thick Concrete Exterior Slab with No Vehicular Traffic and the Slab Extending Not Less than 6 Inches Beyond the Underground Installation	18	4	4	6 (Direct Burial) 4 (In Raceway)	6 (Direct Burial) 4 (In Raceway)
Under Streets, Highways, Roads, Alleys, Driveways, and Parking Lots	24	24	24	24	24
One- and Two-Family Dwelling Driveways and Outdoor Parking Areas, and Used Only for Dwelling-Related Purposes	18	18	18	12	18
In or Under Airport Runways, Including Adjacent Areas Where Trespassing Prohibited	18	18	18	18	18

Note 1. For SI units: 1 inch = 25.4 mm.
Note 2. Raceways approved for burial only where concrete encased shall require concrete envelope not less than 2 inches thick.
Note 3. Lesser depths shall be permitted where cables and conductors rise for terminations or splices or where access is otherwise required.
Note 4. Where one of the wiring method types listed in columns 1–3 is used for one of the circuit types in columns 4 and 5, the shallower depth of burial shall be permitted.
Note 5. Where solid rock is encountered, all wiring shall be installed in metal or nonmetallic raceway permitted for direct burial. The raceways shall be covered by a minimum of 2 inches of concrete extending down to rock.

Figure 15–5 Burial depth requirements for 600 volts and less installations. (Reprinted with permission from NFPA 70–1996, *National Electrical Code*®, Copyright © 1995, National Fire Protection Association, Quincy, Massachusetts 02269. This reprinted material is not the complete and official position of the NFPA on the referenced subject, which is represented only by the standard in its entirety)

Table 710-4(b). Minimum Cover Requirements (Cover is defined as the shortest distance in inches measured between a point on the top surface of any direct buried conductor, cable, conduit, or other raceway and the top surface of finished grade, concrete, or similar cover).

Circuit Voltage	Direct Buried Cables	Rigid Nonmetallic Conduit Approved for Direct Burial*	Rigid Metal Conduit and Intermediate Metal Conduit
Over 600–22kV	30	18	6
Over 22kV–40kV	36	24	6
Over 40kV	42	30	6

For SI units: 1 inch = 25.4 mm.
*Listed by a qualified testing agency as suitable for direct burial without encasement. All other nonmetallic systems shall require 2 inches (50.8 mm) of concrete or equivalent above conduit in addition to above depth.

Exception No. 1: Areas subject to vehicular traffic, such as thoroughfares or commercial parking areas, shall have a minimum cover of 24 inches (610 mm).

Exception No. 2: The minimum cover requirements for other than rigid metal conduit and intermediate metal conduit shall be permitted to be reduced 6 inches (152 mm) for each 2 inches (50.8 mm) of concrete or equivalent protection placed in the trench over the underground installation.

Exception No. 3: The minimum cover requirements shall not apply to conduits or other raceways that are located under a building or exterior concrete slab not less than 4 inches (102 mm) in thickness and extending not less than 6 inches (152 mm) beyond the underground installation. A warning ribbon or other effective means suitable for the conditions shall be placed above the underground installation.

Exception No. 4: Lesser depths shall be permitted where cables and conductors rise for terminations or splices or where access is otherwise required.

Exception No. 5: In airport runways, including adjacent defined areas where trespass is prohibited, cable shall be permitted to be buried not less than 18 inches (457 mm) deep and without raceways, concrete enclosure, or equivalent.

Exception No. 6: Raceways installed in solid rock shall be permitted to be buried at lesser depth where covered by 2 inches (50.8 mm) of concrete, which shall be permitted to extend to the rock surface.

Figure 15–6 Burial depth requirements for over 600 volt installations. Note: Local utilities or the authority having jurisdiction requirements may exceed these minimum requirements. (Reprinted with permission from NFPA 70–1996, *National Electrical Code*®, Copyright © 1995, National Fire Protection Association, Quincy, Massachusetts 02269. This reprinted material is not the complete and official position of the NFPA on the referenced subject, which is represented only by the standard in its entirety)

Note: The coefficient of expansion is for the material; therefore, the formulas apply to all types of PVC raceway products. All types of steel raceways and all types of aluminum raceway products are covered by the respective formula applicable. Expansion fittings are not generally required except in extremely long runs where exposed to a high degree of temperature change, such as rooftops.

The designer and installer generally need not be concerned with thermal expansion and contraction in steel and aluminum installations because of the low coefficient of linear expansion. *NEC*® Section 347–9 states an expansion coupling is needed wherever the change in length due to temperature variation will exceed .25 inch *NEC*® Section 300–7(b) states "Raceways shall be provided with expansion joints where necessary to compensate for thermal expansion and contraction." Expansion must be a consideration in all raceways, both metallic and nonmetallic. However, due to the low coefficient of expansion in steel raceways, only about 5% of (PVC) nonmetallic rigid conduit, expansion joints are rarely necessary to comply with *NEC*® Section 300–7(b). However, expansion joints are often necessary in all raceway types, both metallic and nonmetallic, and ferrous and nonferrous for architectural considerations. Where electrical raceway systems cross building expansion joints, which are commonly found in large buildings, the electrical raceway system

Figure 15-7 Seals to prevent moisture from condensation as required per *NEC*® Section 300-7 where conduit systems pass from areas with widely different temperatures. This section does not require an approved hazardous area seal fitting.

should include an expansion joint to permit the raceway to move with the building. Other considerations for expansion may be encountered where buildings are designed as earthquake proof, such as commonly found in the West Coast areas. Where these flexible architectural joints are encountered, the electrical raceway system should also include expansion joints to compensate for any movement encountered during an earthquake (see Figures 15-8 and 15-9).

Nonmetallic raceways, however, due to the high coefficient of expansion, have special requirements in *NEC*® Article 347. *NEC*® Section 347-9 states that expansion joints for rigid nonmetallic conduit shall be provided to compensate for thermal expansion and contraction except where less than .25 inch of expansion is encountered. This new exception in the 1996 *NEC*® clarifies that all installations of rigid nonmetallic conduit encountering an expansion of over .25 inch is required to have an expansion joint installed.

NEC® Chapter 9, Table 10 provides information to the designer and installer as to when expansion is needed (see Figure 15-8). A good rule of thumb in utilizing this table for other than PVC rigid nonmetallic conduit is to apply a multiplier for other materials. A

Table 10. Expansion Characteristics of PVC Rigid Nonmetallic Conduit
Coefficient of Thermal Expansion = 3.38×10^{-5} in./m./°F

Temperature Change in Degrees F	Length Change in Inches per 100 Feet of PVC Conduit	Temperature Change in Degrees F	Length Change in Inches per 100 Feet of PVC Conduit	Temperature Change in Degrees F	Length Change in Inches per 100 Feet of PVC Conduit	Temperature Change in Degrees F	Length Change in Inches per 100 Feet of PVC Conduit
5	0.2	55	2.2	105	4.2	155	6.3
10	0.4	60	2.4	110	4.5	160	6.5
15	0.6	65	2.6	115	4.7	165	6.7
20	0.8	70	2.8	120	4.9	170	6.9
25	1.0	75	3.0	125	5.1	175	7.1
30	1.2	80	3.2	130	5.3	180	7.3
35	1.4	85	3.4	135	5.5	185	7.5
40	1.6	90	3.6	140	5.7	190	7.7
45	1.8	95	3.8	145	5.9	195	7.9
50	2.0	100	4.1	150	6.1	200	8.1

Figure 15-8 *NEC*® Chapter 9 Table 10. Rule of thumb: To utilize this table for metal raceways, multiply the calculated length change in the table by 0.10 for aluminum, and 0.05 for steel raceways. (Reprinted with permission from NFPA 70-1996, *National Electrical Code*®, Copyright © 1995, National Fire Protection Association, Quincy, Massachusetts 02269. This reprinted material is not the complete and official position of the NFPA on the referenced subject, which is represented only by the standard in its entirety)

Example

380 ft. of conduit is to be installed on the outside of a building exposed to the sun in a single straight run. It is expected that the conduit will vary in temperature from 0°F in the winter to 140°F in the summer (this includes the 30°F for radiant heating from the sun). The installation is to be made at a conduit temperature of 90°F. From the table, a 140°F temperature change will cause a 5.7 in. length change in 100 ft. of conduit. The total change for this example is 5.7″ × 3.8 = 21.67″ which should be rounded to 22″. The number of expansion couplings will be 22 ÷ coupling range (6″ for E945, 2″ for E955). If the E945 coupling is used, the number will be 22 ÷ 6 = 3.67 which should be rounded to 4. The coupling should be placed at 95 ft. intervals (380 ÷ 4). The proper piston setting at the time of installation is calculated as explained above.

$$0 = \left[\frac{140 - 90}{140} \right] 6.0 = 2.1 \text{ in.}$$

Insert the piston into the barrel to the maximum depth. Place a mark on the piston at the end of the barrel. To properly set the piston, pull the piston out of the barrel to correspond to the 2.1 in. calculated above.

See the drawing below.

Figure 15–9 Calculations above are for rigid nonmetallic conduit. For steel raceways, multiply the calculated length change of 21.7 inches by 0.05 which equals 1.08 inches of expansion. For aluminum raceways, multiply the calculated length change of 21.7 inches by 0.10 equals 2.17 inches of expansion. (Courtesy of Carlon, a Lampson Sessions Company)

multiplier of .05 would result in the projected expansion of ferrous (steel) metal conduit, or about 1/20 of PVC. A multiplier of .1 would result in the projected expansion of (aluminum) nonferrous rigid metal conduit, or about 1/10 of PVC. To clarify, see the example in Figure 15–9.

A manufacturer of nonmetallic conduit recommends that 30°F be added to the estimated temperature range when PVC conduit is installed in direct sunlight to allow for radiant heating. This is due to the gray color of the raceway which absorbs that radiant heating.

An expansion coupling consists of two sections of conduit, one telescoping inside another. When installing expansion couplings, alignment of the piston and the barrel is important. Be sure to mount the expansion joint level or plumb for best performance. For a vertical run, the expansion coupling must be installed close to the top of the run with the barrel portion of the expansion fitting pointing down so that rainwater does not run into the opening. The lower end of the conduit run must be secured at the bottom so that any length change due to temperature variation will result in an upward movement (see Figure 15–10).

Example: 380 feet of nonmetallic conduit is to be installed on the outside of a building exposed to the sun in a single straight run. It is expected that the conduit will vary in temperature from 0°F in the winter to 100°F in the summer. We must add 30°F for radiant heating from the sun for a total of 140°F. The installation is to be made at a conduit temperature of 90°F.

- From the table, a 140°F temperature change will cause a 5.7 inch length change for each 100 feet of conduit.
- Thus, the total change for this example is 5.7 inches × 3.8 = 21.67 inches (which should be rounded up to 22 inches).
- Expansion couplings come in two types; one has a range of 2 inches, the other a range of 6 inches.
- The number of expansion couplings will be 22 divided by the coupling range (6 inches or 2 inches). *NEC*® 300–15(c) requires that all fittings and connectors be used only with specific wiring methods for which they are designed and listed.
- If the 6-inch expansion coupling is used, the number will be 22 divided by 6 = 3.67 (which must be rounded up to 4 expansion couplings).
- The couplings should be placed at 95-foot intervals (380 feet divided by 4).
- The proper piston setting at the time of installation is calculated as explained in Figure 15–10. Formula X, 0 = (140-90/140) 6.0 = 2.1 inch.
- Insert the piston into the barrel to the maximum depth; place a mark on the piston at the end of the barrel to properly set the piston; pull the piston out of the barrel to correspond to the 2.1 inches calculated above.
- If this same 380-foot run of conduit was installed using steel conduit, the expansion would be 21.67 × .05, or approximately 1.1 inches. If run in aluminum conduit, expansion would be 21.67 × .1 or approximately 2.2 inches.

Summary

1. Anticipate expansion and contraction of PVC conduit in all aboveground exposed installations.
2. Use an expansion coupling when the length change due to temperature variation will exceed .25 inches.

Determine The Piston Opening

The expansion joint must be installed to allow both expansion and contraction of the conduit run. The correct piston opening for any installation condition should use the following formula:

$$0 = \left[\frac{T\max - T\text{ installed}}{\Delta T} \right] E$$

where:
- 0 = piston opening (in.)
- T max = maximum anticipated temperature of conduit (°F)
- T ins = temperature of conduit at time of installation (°F)
- ΔT = total change in temperature of conduit (°F)
- E = expansion allowance built into each expansion coupling (in.)

STANDARD EXPANSION COUPLINGS

E945 series expansion couplings are designed to compensate for length changes due to temperature variations in exposed conduit runs. See explanation on previous page.

NOTE: Do not use expansion coupling when encased in concrete. Conduit is immobilized by the concrete and will conform to the expansion rate of the concrete.

Expansion Couplings

Part No.	Size	Ctn. Qty.	Lay Lengths Stop to Stop Total Closed	Lay Lengths Stop to Stop Total Open	Available Length Expansion-Contraction
E945D	1/2	50	12 1/4	18 5/8	6
E945E	3/4	50	12 1/4	18 5/8	6
E945F	1	45	12 3/4	19 1/8	6
E945G	1 1/4	30	12 3/4	19 1/8	6
E945H	1 1/2	25	12 3/4	19 1/8	6
E945J	2	15	13 1/2	19 7/8	6
E945K	2 1/2	10	14	20 3/8	6
E945L	3	10	16 1/2	23	6
E945M	3 1/2	5	16 1/2	23	6
E945N	4	5	17 1/2	24	6
E945P	5	3	18 1/2	24 1/2	6
E945R	6	2	20 1/2	26 1/2	6

SHORT EXPANSION COUPLINGS
(Expands to a maximum of 2″)

Part No.	Size	Ctn. Qty.
E955D	1/2	40
E955E	3/4	40
E955F	1	25
E955G	1 1/4	15
E955H	1 1/2	10
E955J	2	6

Figure 15–10 (Courtesy of Carlon, a Lampson Sessions Company)

3. PVC conduit expands 4.1 inches for each 100 feet run and 100°F of temperature change.
4. Steel conduit expands approximately .2 inches per each 100 foot run and 100°F temperature change.
5. Aluminum conduit would expand approximately .4 inches for each 100 feet run and 100°F temperature change.
6. Align expansion coupling with conduit run to prevent binding (plumb or level).
7. Follow the instructions to set the piston opening.
8. Rigidly fix the outer barrel of the expansion coupling so it cannot move.
9. Mount the conduit connected to the piston loose enough to allow the conduit to move as the temperature changes (see Figures 15–8, 15–9, and 15–10).

Note: Manufacturer recommends that for all above-ground installations where temperature change in excess of 25°F (15°C) is anticipated, expansion joints shall be installed. See *NEC®* Chapter 9, Table 10 for expansion characteristics.

SECURE AND SUPPORT REQUIREMENTS

Although the specific requirements for securing and supporting each wiring method type is covered in the article governing that wiring method, general securing and supporting requirements are covered in Article 300. (All support methods are made up from hardware items which are not "listed" by an approved testing laboratory.) The integrity of support methods must be field-evaluated (see Figure 15–11). Careful consideration should be given to the methods used since the integrity of the wiring method may be jeopardized by the use of inadequate support methods. *NEC®* Section 300–11, requires that all raceways, cable assemblies, boxes, cabinets, and fittings be supported and securely fastened in place. Ceiling support wires that do not provide secure support are not permitted as the sole support. This *NEC®* section permits branch circuit raceways to be supported on support wires where the equipment is supported by or located below the suspended ceiling, but are not permitted to be used when used for the support of fire-rated ceilings.

Figure 15–11 Rigid junction box support as well as individual raceway support that must be approved by the AHJ. (Courtesy of Allied Tube & Conduit)

It is important to remember that Section 30–11 only covers branch circuits. *NEC®* Sections 725–6, 760–7, 770–7, 800–5, and 820–5 require access to electrical equipment behind panels, including suspended ceilings which are designed to allow access. These sections further state that access shall not be denied by an accumulation of wires and cables that prevent access.

Caution: The designer should coordinate this support system with the architect designing the ceiling grid system.

Building inspectors may not permit the use of these support wires and may require independent support of raceway systems. Raceways are not to be used as a means of support for other raceways, cables, or nonelectrical equipment, except where identified for the purpose. Two additional exceptions to this rule can be found in Section 300–11(b). Good support methods are important in maintaining the integrity of the raceway system. Generally each wiring component must be independently supported.

As an example of supporting requirements, each component part of the system must be independently supported and in compliance with Article 300 generally.

- The outlet or junction box must be supported in accordance with *NEC®* Article 370.
- The wiring method must be supported in accordance with the article governing it, such as electrical metallic tubing (EMT) Article 348 (refer to Figures 6–3 and 7–4 in chapters 6 and 7).
- The lighting fixtures must be in accordance with Article 410 (see Figure 15–12).

MECHANICAL AND ELECTRICAL CONTINUITY

Metal or nonmetallic raceways, cables, armors, and cable sheaths shall be continuous between cabinets,

Figure 15–12 Strut may serve as a raceway in accordance with Article 352 Part C, or as a support or both. (Courtesy of Allied Tube & Conduit)

boxes, fittings, and other enclosures or outlets. There is an exception to this requirement in Section 300–12. The electrical continuity of conductors in raceways shall be continuous, and there shall be no splice or tap within the raceway itself. Again, there are some exceptions to this rule where it is desirable to allow splices and taps within the raceway, fitting, or box. These include auxiliary gutters, wireways, boxes, fittings, surface metal raceways, and bus ways.

> **Note:** Wrench-tight fittings and couplings, and the proper supporting of the wiring method will insure the compliance of these requirements.

INDUCTION HEATING

When an alternating current is flowing, an electromagnetic field exists around each energized conductor. It varies in strength as the current in the conductor varies from zero to maximum during each half-cycle. If the conductors in a circuit are in separate steel raceways, the changing electromagnetic field induces an electromotive force in each raceway. Current will then flow in each raceway. The magnitude of the current flow is determined by the impedance of the raceway as a current path. Current flow through this impedance may generate sufficient heat to raise the temperature of the raceway high enough to damage the conductor insulation. The current may also cause an additional voltage drop by inducing a back electromotive force in the conductor itself. One solution is to have a conductor of each phase and the neutral of the circuit and equipment grounding conductor, where required, in each raceway or cable. The sum of the currents in one direction would equal the sum of the currents in the opposite direction at any instant, and the changing electromagnetic fields will balance by canceling each other out. Inductive heating will therefore be minimized (see Figure 15–13). *NEC®* Section 300–20 requires electrical circuits enclosed in metal raceways or cables to have each phase conductor, neutral, and equipment grounding conductor, where used, installed in them. This same requirement applies to parallel conductor circuits carried over two or more parallel metal raceways or cables. Inductive heating may occur where the conductors of the circuit pass through the individual openings in the wall of the wireway or auxiliary gutter, metal pole box, or cabinet.

Induced Currents in Steel Raceways

NEC Sec. 300-20
Unbalanced load, 3-phase, 4-wire system

Entire loop subject to inductive heating (violates *NEC Section 300-20*).

Neutral and phase conductors in each raceway.

Figure 15–13 Example of an installation that is in violation of Section 300–20(a) and (b) and will cause induced heating and fail (top), and a correct installation (bottom). (Courtesy of American Iron & Steel Institute)

When feeders are terminated at a panelboard, the inductive heating effect can be minimized by cutting slots in the metal between the conductor openings. Another method is to cover a rectangular opening cut in the metal cabinet with an insulated block containing individual openings for separate conductors. One or the other of these methods is required by *NEC®* Section 300–20.

There is an exception to this requirement where current is negligible, such as for secondary conductors of X-ray and electric discharge sign circuits. Aluminum is not a magnetic metal; therefore, heating due to hysteresis will not occur. However, induced current will be present. The heating will not be of sufficient magnitude to require the grouping of conductors or special treatment in passing the conductors through aluminum wall sections.

NEC® Section 300–3(b) also requires all of the conductors of the same circuit, and where used, the neutral and all equipment-grounding conductors be contained within the same raceway, cable tray, trench, cable, or cord, with some exceptions. *NEC®* Section 300–5(a) has similar requirements for direct buried circuitry in a raceway or cable, and they shall be installed in close proximity in the same trench where there are direct buried cables. Exception 2 of *NEC* Section 300–5(i) permits isolated phase installations in nonmetallic raceways in close proximity where the conductors are paralleled as permitted in *NEC* Section 310–4, and the conditions of *NEC* Section 300–20 are met. This is not a recommended practice, however. *NEC* Section 310–4 makes very clear the importance of paralleling conductors to prevent imbalance loading or induction by stating that the conductors must be the same length, have the same conductor material, be the same size in circular mil area, have the same insulation type, and be terminated in the same manner.

> Fine print note (FPN) to *NEC* Section 310–4 alerts us that the differences in inductive reactance and the unequal division of current can be minimized by choice of materials, methods of construction, and the orientation of the conductors. (It is not the intent to require that the conductors of ungrounded (one-phase), neutral, or grounded circuit conductor be the same as those of another phase, neutral, or grounded circuit conductor to achieve balance (see Figure 15–13).

RESISTANCE TO DETERIORATION FROM CORROSION AND REAGENTS

> **Warning:** Chapter 5 of the *NEC* does not consider the effects of deterioration from corrosion and reagents. It is the responsibility of the designer to consider the environment of the hazardous classified location.

NEC Section 110–3 requires the designer and installer to select the proper materials for each installation (see Figure 15–14). Guidelines for making these selections can be found in Underwriters Laboratories standards, the UL "Green Book," the *Electrical Constructions Materials Directory*, or the UL "White Book," which is the *General Information Directory for Electrical Construction, Hazardous Location and Electrical Heating and Air Conditioning Equipment.*

Figure 15–14 Examples of PVC-coated galvanized steel rigid conduit which offers excellent protection from all types of corrosion and physical protection. (Courtesy of Robroy Industries)

In addition, *NEC* Section 110–11 reminds us that deteriorating agents contribute to corrosion. Unless identified for use in the operating environment, no conductors or equipment shall be located in that environment. A new (FPN) fine print note number two was added in the 1996 *NEC* Section 110–11 which states "Some cleaning and lubricating compounds can cause severe deterioration of many plastic materials used for insulating and structural applications in equipment." *NEC* Section 300–6 reminds us that metal raceways, armored cable, boxes, cable sheathing, cabinets, elbows, couplings, fittings, supports, and support hardware shall be of material suitable for the environment in which they are installed. Outer coatings should be of an approved corrosion resistant material, such as zinc, cadmium, or enamel (see Figures 15–15 and 15–16). Where protected from corrosion solely by enamel, they shall not be used out-of-doors or in wet locations as described below. Ferrous and nonferrous metal raceways, cable armor, boxes, cable sheathing, cabinets, elbows, couplings, fittings, supports, and support hardware shall be permitted to be installed in concrete or in direct contact with the earth and areas subject to severe corrosive influences when made of material judged suitable for the condition and when provided with corrosion protection approved for the condition (see Figure 15–14). In portions of dairies, laundries, canneries, and such locations where the walls are frequently

Rigid Steel Conduit, Intermediate Steel Conduit (IMC), and Electrical Metallic Tubing (EMT) are *Corrosion Resistant* for Wide Applicability

Both the exterior and the interior of tubular steel raceways are coated to protect from corrosion. Coatings are usually of zinc.

Zinc coating, or galvanizing, effectively seals the steel surfaces from corrosive moisture and other corrosive agents. In addition, the zinc coating may slowly sacrifice itself—and thus save the steel surfaces from corrosion—in those cases where the conduit may have been damaged and the steel exposed by a blow.

There are four methods of galvanizing steel conduit:

1. Hot-Dipping
2. Electroplating
3. Sherardizing
4. Metallizing

In the hop-dipping process, the clean tube is immersed in a molten zinc for a suitable time and then withdrawn. The tube is wiped with a superheated steam or air to remove any excess of zinc on either the exterior or interior, and to control the thickness of zinc.

In the electroplating process, the clean tube is placed in a solution of zinc salts and made a cathode of an electrical circuit so that the zinc is plated from the solution onto the outside of the tubing. Positive contact with the solution is obtained through zinc anodes, which replace the zinc content of the solution as it is used up.

In the sherardizing process, the conduit is first coated with zinc by heating a predetermined amount of commercially pure zinc powder and the steel pipe in a slowly revolving, sealed retort. A second method is also employed of electroplating a controlled thickness of zinc coating to the steel pipe. Both methods of zinc coating are subjected to accurately controlled furnace heat.

In the metallizing process, finely divided particles of zinc are metal-sprayed in a heated, semimolten condition onto the surface of a specially prepared tube to form an adherent coating.

Although hot-dipping and electroplating are the most common, any of the four methods of coating gives the tube a uniform coating of zinc of excellent corrosion resistance. (See A.S.A. for thickness of coating and test procedures.)

Figure 15–15

TABLE OF METALS IN GALVANIC SERIES
Corroded End (anodic or less noble)

MAGNESIUM
ZINC
ALUMINUM
CADMIUM
IRON OR STEEL
TIN
LEAD
NICKEL
BRASS
BRONZES
NICKEL-COPPER
ALLOYS COPPER

Any one of these metals and alloys will theoretically corrode while offering protection to any other that is lower in the series, so long as both are electrically connected.

As a practical matter, however, zinc is by far the most effective in this respect.

Because zinc is more active in galvanic couples than iron and steel, it provides these metals with electrolytic protection against rust. This protection is so effective that even though there be a small exposed area on the base metal, the attack of the elements will be directed to the zinc, and protection will continue as long as sufficient zinc remains.

This is what happens at a small exposed area in a coating of a metal having a lesser tendency to go into solution than the base metal. The steel rusts away, protecting the tin.

This is what happens at a small exposed area in a coating of a metal having a greater tendency to go into solution than the base metal. The zinc is consumed while protecting the steel from any attack.

Position in the galvanic series is not an infallible guide to galvanic protection because of other influences. Zinc is the one metal that will protect iron and steel under conditions of exposure to be expected.

Figure 15–16

washed, or where surfaces are of absorbent materials such as damp paper or wood, the entire wiring system shall be mounted so that there is at least ¼-inch of air space separation from the wall or other mounting structure. In general, areas where acids and alkali chemicals are stored or may be present are highly corrosive areas, particularly when these areas are wet or damp. Severe corrosive conditions may also be present in portions of meat-packing plants, tanneries, glue houses, stables, installations immediately adjacent to the seashore, and swimming pool areas. Areas where chemical de-icers are used and storage cellars or rooms for animal hides or casings, fertilizer, salt, and bulk chemicals are also to be considered highly corrosive.

NEC Article 547 gives us some additional alerts when the installation falls within the scope of agricultural buildings and the concerns for corrosive atmospheres exist. These include poultry and animal facilities that may be damp from washing or sanitizing, where cleaning agents and similar conditions exist, livestock confinement areas, environmentally controlled poultry houses, and similar enclosed areas of an agricultural nature. With these sections, there is ample warning to the designer to make a correct and proper selection of the material to be specified. Since 1905, rigid metal conduit has been manufactured with a corrosion protection, originally a coating of black enamel, later by means of electro-galvanizing and hot-dip galvanizing. In 1961, the first PVC coated rigid steel conduit was offered. Today there are numerous products on the market that provide specialty coatings for specific concerns relating to corrosion and deteriorating factors from the presence of severe environment and chemicals. Underwriters Laboratories, based on their tests and evaluations, state in the *General Information for Electrical Construction, Hazardous Location, and Electric Heating and Air Conditioning Equipment 1991* "White Book" that the intermediate metal conduit, as covered by *NEC* Article 345, and rigid metal conduit, as per *NEC* Article 346, installed in concrete does not require supplementary corrosion protection. Rigid and IMC conduit installed in contact with soil does not generally require supplementary corrosion protection. EMT as covered in Article 348 does not generally require supplementary protection in concrete on grade or above, but generally will if in concrete below grade or in direct contact with the soil. Aluminum EMT in concrete or direct soil contact always requires supplementary protection. In the absence of specific local experience, soils producing severe corrosive effects are generally characterized by a low resistivity of less than 2000 ohm-centimeters. Wherever ferrous metal conduit or tubing runs directly from concrete encasement to soil burial, severe corrosive effects are likely to occur on metal in contact with the soil. Supplementary nonmetallic coatings as part of the conduit have not been investigated for resistance to corrosion. Manufacturers should be contacted for this data. Supplementary nonmetallic coatings of greater than .010-inch thickness applied over metallic protective coatings are investigated with respect to flame propagation and detrimental effects to the basic corrosion protection provided by the protective coatings (see Figure 15–15). Many manufacturers today, in addition to the galvanized coating on metal raceways, also provide an additional layer of chromate or other organic coating to prevent formation of "white rust." "White Rust" is the powdery substance which otherwise forms as a protective process of zinc (see Figure 15–15). This provides some additional protection, which is not evaluated by UL. However, it certainly adds to the appearance as well as the protection on those raceway systems.

Many questions arise as to the suitability of the galvanized protective coating applied to rigid, IMC, and EMT conduit. The zinc coating need not be unscratched to remain effective. If a scratch should expose bare steel, moisture in the air will form an electrolyte on the surface of the raceway and through galvanic action, zinc will be transported to the exposed steel and deposited or plated out as an electrolytic band. This healing action of zinc affords protection even where wrench marks or scratches are as much as 0.1-inch wide. These metallic raceway types have provided excellent serviceability in countless installations over the past 100 years (see Figure 15–15). Corrosion is generally not a problem in most soil types. However, the final decision must be made by the designer and approved by the authority having jurisdiction.

Rigid nonmetallic conduit has been perceived by many installers to be impervious to all corrosive atmospheres. This is not true! Rigid nonmetallic conduit, both Schedule 40 and Schedule 80, are generally acceptable for many environments; however, when making an installation in environments containing chemicals, check the manufacturer's recommendations (see Figure A–13 in the Appendix). If there are any questions for specific suitability in a given environment, prototype samples should be tested under actual conditions. Underwriters Laboratories *General Information from Electrical Construction Materials Directory* "White Book" states listed PVC conduit is inherently resistant to atmospheres containing common industrial corrosive agents and will also withstand vapors or mist from caustic, pickling acids, plating bath, and hydrofluoric and chromic acids. PVC conduit, elbows and bends including couplings which have been investigated for direct exposure to other reagents, may be identified by the designation "Reagent Resistant" printed on the surface of the product. Such special uses are described as follows: "PVC conduit, elbows, and bends where exposed to the following reagents at 60°C or less, acetic, nitric (25°C only) acids, acids in concentrations not exceeding ½ normal; hydrochloric acid in concentrations not exceeding 30 percent; sulfuric acid in concentrations not exceeding 10 normal; sulfuric acid in concentrations not exceeding 80 percent (25°C only); concentrated or dilute ammonium hydroxide; sodium hydroxide solutions in concentrations not exceeding 50 percent; saturated or dilute sodium chloride solution; cottonseed oil, or ASTM #3 Petroleum Oil." An additional alternative was offered in 1961 when the first PVC-coated rigid steel was pioneered. Today there are several different types of coated conduit designed for different corrosive environments. These manufacturers offer a complete wiring method in special environments. They include coated conduit, fittings and accessories, a complete package line, and furnish experienced sales and engineering personnel capable of recommending the proper system for your application and providing a complete bill of material including all conduit fittings, junction boxes, conduit bodies, and hardware for a complete system. Installation is made simple with standard rigid conduit tooling and damage to the surface can easily be repaired by the installer at the time the damage occurs. The manufacturers of these products use standard listed galvanized rigid conduit in conformance with NC C80.1 and Underwriters Laboratories UL6 in accordance with *NEC®* Article 346 or listed intermediate metal conduit in compliance with UL 1242 and NC C80.6 in accordance with *NEC®* Article 345 listed products, which is then coated to NEMA Standard RN–1. That material then has a 15-to-40-mils thick protective coating applied to the outside surface of the conduit. Protection is available on the outside or inside or both in various mil thicknesses, according to the specific needs and is suitable in all hazardous, heavy industrial, manufacturing, corrosive areas, bridges, highways, utility plants, and substations, etc. These products afford excellent protection from corrosion and reagents and deterioration from reagents, as well as excellent physical protection, and are ideal where both these conditions exist, (see Figure 15–14). **Note:** See Interior and Exterior Chemical Resistance Charts in Figure A–15 in the Appendix. Today's designers and installers are offered an excellent selection of materials to meet all the needs of today's state-of-the-art installations.

> **Note:** See Figures A–13 and A–15 (manufacturer corrosion resistance data) in the Appendix.

PENETRATION OF FIRE-RESISTIVE ASSEMBLIES

There are three major building codes which include many provisions, the purpose of which is to provide construction that not only will withstand the effects of fire in a portion of a building, but will also reduce the likelihood of fire spreading within the building. For extremely large or high structures, building codes usually require fire-resistive construction, as well as the use of noncombustible construction materials. Building codes prescribe that in fire-resistive buildings, the columns, the beams, girders, walls, and floor constructions must have a resistance to fire measured by a standard fire test procedure (ASTM E119, Standard Method of Fire Tests of Building Construction and Materials). The specific requirements of the assembly design under the classification system are found in Underwriters Laboratories *Fire Resistive Directory 1991* ("Orange Book") and the *Building Materials Directory* ("Manila Book"). Other accredited labs also conduct tests to establish fire ratings. The three model building codes govern the fire-resistance requirements and stipulate the time-rated period for various types of occupancies (such as 1-hour, 2-hour, etc.). These requirements are found in the BOCA National Building Code Article 9, ICBO Uniform Building Code Chapter 43, and SBCCI Standard Building Code Chapter 10.

Many fire-resistive assemblies use cellular floors, decks, and suspended ceilings. Openings for recessed lighting fixtures, electrical outlets, and air vents may be included in these fire-resistive assemblies. The ceiling usually is, but may not be, tested as part of the assembly. In any event, the details of the construction are important if the required fire resistance is to be achieved. Fire-resistive assemblies tested by UL are "classified." This is the same approval as "listed," but connotative of testing for a very specific application in a specific manner. It is essential that a listed fire-resistive assembly conform to the tested specimen in all particulars after field installation. The materials and dimensions must be at least equal to those of the test specimen and constructed per the classified design. Any holes or apertures in the fire protection must be limited in number, area, and distribution in the same manner as in the test assembly. Engineering analysis also must show that fire resistance of an assembly is not critically affected by a proposed arrangement of openings for lighting fixtures or ducts. The code permits openings for outlets, air-conditioning ducts, and similar equipment where such openings do not exceed the percentage of the ceiling area established by fire test data. Because such openings allow only a small amount of heated gases generated by a fire to enter a concealed space, their hazard is considered negligible. Openings of this type are often described as a *membrane penetration*. Obviously, no such penetration shall extend through an entire wall or floor and ceiling assembly. The design and construction of fire-rated assemblies are quite specific, even including specific nails and screws. Minor changes in design, such as thickness or the method of fastening of construction materials, may have a significant effect on the assembly's behavior during a fire. Wires supporting a suspended grid ceiling, for example, are critical; for this reason, the *NEC®* does not allow such wires to be used for support of electrical wiring when it is a rated ceiling (see *NEC®* Section 300–11[a]). If a wall or ceiling is perforated for the passage of raceways or for work space around them, the fire resistance of the wall or ceiling is effectively destroyed unless the opening around the penetrating item is properly sealed. This danger is recognized in the *NEC®* Section 300–21, which states: "Electrical installations in hollow spaces, vertical shafts, and ventilation or air-handling ducts shall be so made that the possible spread of fire or products of combustion will not be substantially increased. Openings around electrical penetrations through fire-resistance rated walls, partitions, floors, or ceilings shall be firestopped using approved methods to maintain the fire-resistance rating." Firestopping must be installed in accordance with the local adopted building and fire codes.

Because most plumbing, air-handling, electrical, communications, and other building services must pass through fire-resistant assemblies, coordination of building design, type of materials, and installation methods for these systems is essential to minimize the danger of spread of fire. Changes in design of or in materials used in a fire-rated assembly should not be introduced without pre-approval of the authority having jurisdiction.

Plans for proposed buildings may not indicate all eventual penetrations. The absence of detailed provisions for electrical or communications systems in the floor plan does not mean that such systems will not be installed later. If the electrical and communications circuits are to be added after construction is completed, it may be necessary either to construct chases (shafts) in fire-rated floors, walls, and ceilings for cables or raceways to pass to other areas or openings may be made in the walls, floors, or ceilings. Such penetrations through an entire assembly may void the intended fire protection of the construction and allow the spread of any of the products of combustion (smoke, hot or toxic gases, or flame). Creation of passageways through which smoke, gases, and heat may pass only increases the hazards to the building's occupants. It is the responsibility of the authority having jurisdiction to determine whether a particular firestop method or material is to be approved. This is a significant responsibility for the AHJ who usually must look to others for information to verify that a particular method or material has been adequately evaluated.

The building codes provide two recognized methods for sealing openings around penetrating items. These are through-penetration fire-stop systems and annular space filler materials. Through-penetration systems are available for both combustible and noncombustible penetrating items. Annular space fillers may only be used with noncombustible penetrants. Why the difference?

When combustibles like nonmetallic-sheathed cables, insulated conductors in cable trays, and PVC conduit are exposed to heat, they burn or melt away and materials surrounding them fall away, leaving a hole through which the fire could continue to travel either vertically or horizontally to adjoining floors or compartments. In the 1970s a disastrous cable fire spread through the Brown's Ferry Nuclear Plant in just this manner. Because of that event, new methods of sealing were developed; among them are materials which swell up when exposed to heat and fill the hole as the nonmetallic materials burn away. This swelling process is called intumescing. Other through-penetration systems are the endothermic type. This means they release chemically bound molecules of water when exposed to heat. Both types must be tested in accordance with ASTM E815. UL Standard 1579 is the same as ASTM E815, ("Fire Tests of Through-Penetration Firestops"). Through-penetration systems classified by UL are found in the "Orange Book" (*Fire Resistance Directory*). This directory is very explicit about the assembly design, the size and type of the penetrating item(s), the size of the opening around the item, and the thickness and construction of the protection materials. It is very important that the listing criteria be followed precisely as the systems are designed to assure enough volume to fill the opening when the penetrating item burns or melts away or to control the temperature on the unexposed side. UL is explicit that individual components are not rated, nor are they intended to be intermixed among systems. The classified rating applies only to the complete system. There are some who think through-penetration fire-stop systems prevent the passage of smoke and toxic gases. While this may be true in some instances, the E815 test does not provide for determining this; so it is an unknown.

Through-penetration fire-stop systems are absolutely necessary for combustible penetrations. For noncombustible penetrations, like EMT, IMC, and rigid conduit, either through-penetration systems or annular space fillers may be used. The systems for metallic penetrations are also found in the UL "Orange Book." To use this book, for both combustibles and noncombustibles, you will need to know the specifics of the assembly being penetrated, such as the fire-rating, type of studs, thickness of gypsum or concrete, as well as the size and type of penetrating item(s).

The "Orange Book" indicates an "F" rating and a "T" rating. The "F" rating is based upon the time period for which the system prohibits flame passage to the unexposed side (as evidenced by flame occurrence) in conjunction with an acceptable hose stream test performance.

Unless the penetration is contained in a wall cavity you may also need to know the "T" rating given to the specific system. The "T" rating is the test time expired before the sealant and penetrating item reach a temperature of 325° F above ambient on the non-fire side of the assembly in addition to containing the flame and passing the hose stream test. Hose stream testing is the application of a specified water stream from a fire hose to evaluate the structural integrity of the assembly after fire exposure. It is not relative to fire fighting activities during an actual fire. You should consult the applicable building code for exact language regarding "T" ratings as this is still in the process of change and varies in different codes.

Annular space fillers, which are permitted to be used only with noncombustible penetrants, do not require "T" ratings. However, the building codes do require that, to qualify as annular space fillers, materials must be able to withstand ASTM E119 time-temperature conditions under positive pressure for the full period for which an assembly has been rated without igniting cotton waste placed in contact with it.

Common construction materials have long been used with great success as annular space fillers. These are products such as cement, mortar, grout, handyman caulk, mineral wool, and even joint compound. As an aid to the industry the "NEMA Steel Rigid Conduit and EMT Section (5RN)" sponsored an investigation at UL to document the performance of these longtime annular space fillers.

These were generic-type tests and easily affirmed the performance of these materials. Based on this testing, the three national model building codes have added language which automatically accepts cement, mortar, and grout for sealing openings in masonry assemblies. The applicable code should be reviewed for specifics because there are some variables in the precise language.

> A summary (or full) report on this annular space testing is available from NEMA, Section 5RN, 2101 L Street, NW, Suite 300, Washington, D.C. 20037. Ask for "UL Special Services Investigation Fire NC 546 Project 90NK11650."

This report documents that everyday joint compound can maintain the fire-resistance rating of a gypsum wallboard assembly for two hours. The report shows that even with the damage to the gypsum board (½-inch sheetrock) on the fire side of the test specimen, which occurred when the furnace exploded, the fire-resistance rating of two hours was achieved. It is a good document for use in securing AHJ approval when planning to seal with annular space filler.

Electricians and designers must become educated on the sealing of openings around electrical wiring in order to comply with *NEC®* Article 300–21 and with the building codes.

COMBUSTIBILITY OF RACEWAY MATERIALS AND CONDUCTOR INSULATION

In the last twelve to fifteen years a great deal of attention has been given to hazards of fire relative to combustible electrical systems (along with combustible construction furnishings and finishing materials). Much of this attention was created by the expansion in the *NEC®* of nonmetallic wiring methods within buildings. The degree of the hazard remains controversial, as does the risk involved. Certainly, individuals draw different conclusions from the same set of facts; additionally, risk tolerance varies from person to person. Otherwise we would not have deep-sea divers and fighters of oil well fires. Other questions have also arisen in this controversy, such as how many lives is it acceptable to lose—we can't feasibly get to zero. The age-old response then becomes, your family or mine? In this section we look at some of the issues involved. The user, designer, or jurisdiction then has to determine the acceptable level of risk for the particular situation.

The *NEC®* is a minimum code and allows choices in wiring methods. However, in a few instances it has directly addressed the matter of fire concerns. One specific area is in plenums and other environmental air spaces. The whole construction code arena recognizes the dangers inherent in allowing the products of combustion (smoke, heat, toxic gases) to spread throughout a building by way of the air handling system. Materials allowed in such spaces are carefully controlled (although a series of events in recent years is making this control borderline).

Article 300–22 of the *NEC®* controls the electrical system in environmental air spaces. The wiring methods in ducts or plenums used for environmental air (Section 300–22[b]) must be EMT, IMC, rigid metal conduit, flexible metallic tubing, MI cable, or type MC cable employing a smooth or corrugated impervious metal sheath without an overall nonmetallic covering. In order to connect physically adjustable equipment and devices permitted in these areas, lengths of flexible metal conduit or liquidtight flexible metal conduit not over 4 feet are permitted.

A plenum is defined in the *NEC®* as a "compartment or chamber to which one or more air ducts are connected and which forms part of the air distribution system." The BOCA National Building Code contains the definition: "An enclosed portion of the building structure which forms part of an air distribution system and is designed to allow the movement of air."

Thus, it is important to recognize that a ceiling space through which the building air moves is considered a plenum. If there is ever any doubt relative to the *NEC®* definition, remember that the adopted building code governs.

In environmental air spaces, other than ducts or plenums (Section 300–22[c]), other wiring methods are added to those named above. These include totally enclosed nonventilated, insulated busways having no provisions for plug-in connections, type AC cable, and other factory-assembled multiconductor control or power cable which is specifically listed for the use. Some of these are called "plenum cable," which is a misnomer because they are not for use in actual plenums. One other means of enclosing the conductors, beyond those conduits named in Section 300–22(b), is surface metal raceway or wireway with metal covers or solid bottom metal cable tray with solid metal covers. For minor exceptions, refer to Section 300–22(c).

What are some of the concerns regarding combustible wiring methods, whether or not in environmental air spaces? The primary nonmetallic materials used in buildings for raceways and conductor insulation are PVC and nylon. When PVC reaches a temperature in the range of 450° to 500°F, it starts to chemically break down (thermally decompose) although some formulations may not ignite until almost 1000°F. This decomposition releases a very corrosive and irritating hydrogen chloride gas (HCL) and at this point may not even produce visible smoke. This HCL is very irritating to the eyes, nose, and lungs of those who are exposed to it. As the decomposition progresses, thick black smoke is also produced; actual burning (flaming) is not required. Between the smoke obscuring the path of escape, the burning eyes, and choking sensation, it becomes very difficult to evacuate the area. Even if one is able to get out, sufficient exposure to HCL as a product of combustion (attached to soot particles) can produce severe respiratory problems which are frequently irreversible and at times lead to delayed death. (In toxicity tests, many animal deaths from PVC combustion do not occur until several days later.) PVC also produces large amounts of carbon monoxide. Although nonplasticized PVC, such as that used in raceways, is ignition resistant, it will burn once the chlorine and other fire retardants are driven off. Plasticized PVC, used in conductor insulation, burns more quickly. Another gas produced which is of concern is benzene, which is both flammable and carcinogenic. Burning nylon produces cyanide. Testing at the University of Pittsburgh concluded that 18 to 30 inches of PVC conduit burned in an average room is sufficient to create a dangerous situation and will kill humans in 10 to 15 minutes if decomposed between 300°C (572°F) and 500°C (932°F).

An additional effect of HCL is its corrosive effect on electrical contacts and other electrical and electronic equipment. Deposits left by burning PVC in equipment rooms, even after thorough scrubbing, have been known to cause corrosion weeks and months later. Computer equipment exposed to PVC combustion products during testing has been damaged. The corrosive effects on electrical contacts of HCL from overheated conductors could affect operation of safety and signaling equipment. Testing on the nonmetallic side of the issue shows that HCL from PVC adheres to walls, floors, and room objects. The question is how much of the HCL is therefore removed from the atmosphere—how much under the great variety of fire conditions is the unanswered question. Certain *NEC®* articles also require that nonmetallic conduit be placed behind a thermal barrier to inhibit its exposure to heat for a projected 15 minutes. In such instances it must be assured that inadvertent openings are not made and left unsealed, or that thermal barrier ceiling tiles are not removed.

Noncombustible raceways have long been the recognized and proven method of inhibiting fire spread by way of the electrical system. Even if the conductor insulation inside burns, the spread will be limited; the smoke will be more contained; and the raceway itself will not contribute smoke or hazardous gases.

It is incumbent on users to evaluate the specific set of conditions for an electrical system, including the number and type (as related to their ability to exit in an emergency) of occupants that might be anticipated in a building and its design, and thus determine the level of risk they wish to take relative to combustible materials. It is important to remember that state supreme courts have found that compliance with adopted codes alone does not necessarily preclude fire-injury liability. This does not apply just to the electrical system, but is a consideration for all design.

WIRING IN DUCTS, PLENUMS, AND OTHER AIR-HANDLING SPACES

The provisions for installation of electrical wiring and equipment in ducts, plenums, and other air-handling spaces are covered in *NEC* Article 300, Section 300–22, generally. It should be noted, however, that often building codes and local fire codes take precedence and are usually more restrictive. The *NEC* Article 100 defines a plenum.

> "plenum": *"A compartment or chamber to which one or more air ducts are connected and which forms part of the air distribution system."*

It is advised that the designer or installer check with the chief building official for guidance and direction before encroaching on these areas. No wiring of systems of any type shall be installed in ducts used to transport dust, loose stock, or flammable vapors. The *NEC* Section 300–22(a) restricts wiring of any type from being installed in any duct or shaft containing only such ducts used for vapor removal or for ventilation of commercial-type cooking equipment. Wiring methods approved for ducts or plenums used for environmental air are type MI cable, type MC cable employing a smooth or corrugated impervious metal sheath, electrical metallic tubing, flexible metal conduit, intermediate metal conduit, or rigid metal conduit and they are permitted to be installed in ducts or plenums specifically fabricated to transport environmental air. Flexible metal conduit and liquid-type flexible metal conduit not to exceed 4 feet is permitted where adjustable equipment and devices are located in these chambers. The connectors used for flexible metal conduits shall close openings in the connection. Equipment and devices shall be permitted within such ducts and plenum chambers only where necessary for the direct action or sensing of the contained air. Where such equipment or devices are installed and illumination is necessary to facilitate maintenance and repair, only enclosed gasket-type fixtures are permitted. In accordance with Section 300–22(c), the other space used for environmental air, which is used for purposes other than ducts and plenums is described in Section 300–22(b). An example would be the space such as a space over a suspended ceiling, which is used for environmental air handling purposes. The wiring methods approved in these areas are totally enclosed nonventilated insulated busways having no provision for plug-in connections, wiring methods consisting of MI cable, type MC cable without an overall nonmetallic covering, type AC cable, or other factory assembly multiconductor control or power cable which is specifically listed for the use. Other type cables and conductors shall be installed in electrical metallic tubing, flexible metallic tubing, intermediate metal conduit, rigid metal conduit, flexible metal conduit, or where accessible, surface metal raceway or wireway with metal covers or solid bottom metal cable tray with solid metal covers. Electrical equipment with a metal enclosure or with a nonmetallic enclosure listed for use and having adequate fire resistance and low smoke-producing characteristics and associated wiring materials suitable for the ambient temperature shall be permitted in such areas, unless specifically prohibited in other sections in the *National Electrical Code*. There are limited exceptions to these rules. Data processing raised floor areas are covered in *NEC* Article 645, which requires that all of the conditions in Section 645–2 be met before the area can be classified as a data processing room. The requirements as noted in *NEC* Section 645–2 are that disconnecting means must be installed to disconnect all electronic equipment in the room and all power to dedicated HVAC heating, ventilating, and air-conditioning equipment serving the room, and cause all required fire and smoke dampers to close. The disconnects must be grouped, identified, and controlled from locations readily accessible at all principal exit doors. In addition, if a separate heating, ventilating, and air-conditioning (HVAC) system is provided that is dedicated to the electronic computer and data processing equipment's use and is separated from other areas of occupancy, except other HVAC equipment, it may serve these areas provided fire/smoke dampers are installed at the point of penetration of the room boundary. Such dampers must operate on the activation of smoke detectors; also a disconnecting means is required. Only listed electronic computer and data processing equipment is installed in the room. The room is occupied only by those personnel needed for maintenance and the functional operation of that room, and the room is separated from other occupancies within that building or floor by fire-resistant rated walls, floors, and ceilings, with protected openings. The building construction shall comply with the applicable building codes. Good engineering design and close coordination with the authority having jurisdiction responsible for these occupancies, such as the chief building official, fire marshal, and the chief electrical inspector, should all be consulted before applying Article 645 to an electrical system. Remember that *NEC®* Article 645 permits a more lenient wiring method; therefore, greater caution must be taken to confine a fire to that area and to insure that it does not propagate (spread) to the rest of the building. Steps should be taken to ensure easy egress for those working in that room in the event of an emergency.

RACEWAY ALLOWABLE CONDUCTOR FILL

The allowable conductor fill requirements are found in Chapter 9, the Tables including the Notes to those Tables, and Appendix C of the *NEC®*. With all types of conductors except lead-covered, three conductors or more in a raceway is limited to 40 percent of the cross section of the conduit or tubing. Other allowable percentages of fill are listed in Table 1 of Chapter 9. Equipment grounding or bonding conductors when installed must be included in the calculations. The actual dimensions shall be used in making the calculation whether these conductors are insulated or bare. Conduit nipples not exceeding 24 inches shall be permitted to be filled to 60 percent of the total cross-sectional area. Conductors not included in Chapter 9 shall be based on their actual dimensions for the purpose of making the calculations. Conductors' dimensions are found in *NEC®* Table 8 of Chapter 9. Multiconductor cable of two or more conductors shall be treated as a single conductor when calculating the percentage of conduit fill. For cables that have an elliptical cross section, the calculations shall be based using the major diameter of the ellipse as the circle diameter. The trade size of the raceway in *NEC®* Table 4, Chapter 9 shall be used as the internal diameter in inches except Schedule 80, rigid nonmetallic conduit, in which Table 4A shall be used.

> Example: What size raceway is needed to enclose nine number 12 THW conductors in an intermediate metal conduit? Answer: ½-inch, by contrast electrical nonmetallic tubing would require a ¾-inch. Reference: *NEC®* Chapter 9 Table 1 and Appendix C Table C2.

> Example: What size raceway is needed to enclose four 500 kcmil THWN, one 300 bare conductor? Answer: The 1996 *NEC®* has revised all of the tables in Chapter 9, as a result the type of raceway must be known; for example, EMT the minimum size would be 3-inch, by contrast a Schedule 80 PVC conduit the minimum size would be a 3½-inch. Reference: *NEC®* Chapter 9 Tables 1, 4, 5, and 8.

It should be noted that although the raceway is sized according to the allowable conductor fill in Chapter 9 of the *NEC®*, there are other considerations necessary in sizing the conductors before the raceway system itself is sized, such as the ampacity adjustment factors in Note 8 of the 0–2000 Volt Tables in Article 310 for more than three conductors per raceway. There are ambient temperature considerations, which require ampacity correction where the ambient temperatures exceed levels as listed at the bottom of each of the ampacity tables as they apply. Good engineering practice, however, recommends that when all of these things are considered, the raceway system should be oversized by 50 percent to handle future expansion and growth to the electrical system.

RACEWAY SYSTEM

About a hundred years ago, as we entered the 1900s, our great electrical industry was growing like wildfire. This was a result of the great inventors such as Thomas Edison, holder of hundreds of patents promoting DC voltage; Nikola Tesla, inventor of the 3-phase motor and a promoter of AC voltage; William Merrill,

Figure 15–17 Early inventors searched for the optimum raceway to carry electrical conductors. They found tubular steel conduit to be that raceway. Rigid steel conduit has provided that need for more than one hundred years.

founder of Underwriters Laboratories, along with George Westinghouse, and others, (see Figure 15–17). As a result of this rapid growth in the use of electricity, there were many fires, electrocutions, and electrical accidents almost everywhere. With arcing and sparking and resultant fires occurring almost daily and with keen competition between Edison (DC) and Tesla (AC), a group got together in 1881 to discuss safer use of electricity. The National Association of Fire Engineers met in Richmond, Virginia, and from this meeting came a proposal that served as a basis for the first *National Electrical Code®*. In the earliest beginnings of the electrical era, before the turn of the century, men began to design various means of distributing the wonderful servant made practical by the work of these great inventors—electricity. Ordinary electrical conductors strung without protection soon proved faulty, dangerous, and wasteful. Some means by which they could be enclosed and protected permanently against mechanical and electrical damage was clearly needed. The black-painted steel pipe which originally ran through buildings to provide gas for lighting was already in place and provided the first protection for electrical conductors. From that evolved today's galvanized rigid steel conduit and later the lighter-weight steel electrical metallic tubing as wiring became more widespread. The earlier conduits were of zinc with copper elbows; then spiral-wrapped paper tubing with brass joining sleeves was introduced. As the shortcomings of these conduits were revealed, brass-enclosed paper tubing with brass elbows and couplings was introduced. But this, too, was found to be unsatisfactory, chiefly because of inadequate physical protection and mechanical strength. The concept of a conduit into which electrical conductors could be pulled in and out remained as worthwhile, but the sole question to be answered was what type of material and processing would provide the best and most economical conduit?

Iron-armored conduit was introduced in 1894. This consisted of ten-foot lengths of standard-weight wrought iron gas pipe, with paper tube lining and threaded couplings and nipples. This conduit proved to be a great advance, but was found to be still far from satisfactory in every respect. The linings reduced the usable interior area to an unsatisfactory degree. Linings of paper or fiber would crumble, and wooden linings would splinter, making field bending impossible. These linings were not moisture-proof and were expensive.

The introduction of good rubber-insulated electrical conductors in 1897 made possible the development of unlined steel conduit. Here at last was a conduit that provided full mechanical and electrical protection for conductors. The system could be readily expanded and furnished maximum design flexibility, as well as a host of other desirable features. Conduit also serves as an excellent equipment-grounding conductor. The electrical system could be changed by pulling in new conductors without damaging the walls, floors, and ceilings. These desirable features in one easily installed convenient raceway system led to the widespread use of steel conduit. Although the quick acceptance of this new steel conduit by the electrical industry was followed by rapid improvements in the product, it was not without its problems. The gas fitters immediately claimed the right to install their wrought iron gas pipes, and there ensued great craft disputes between the electrical craftsmen and the gasfitter craftsmen. This was finally resolved, and the electrician was allowed to install conduit. About this same time, groups began to cut wrought iron gas pipe into ten-foot lengths and bootleg it over the counters to the electricians as conduit. The problem was that electrical conduit was rougher on the exterior and smoother on the interior, whereas wrought iron gas pipe was just the opposite—it had a smoother exterior with a rougher interior. The conduit interior was important so as not to tear the rubber insulation on the conductors as they were being pulled in. Fortunately, with the rapid advance in the industry and improvements in the product, these problems were soon far behind.

The first exterior coating to prevent oxidation was enamel. In 1902 the first electro-galvanized rigid steel conduit was introduced. In 1908 sherardized steel conduit appeared, followed by hot-dip galvanizing in 1912. A little later, the metallizing process was introduced.

ADVANTAGES OF RACEWAY SYSTEMS

Since 1905 the two primary functions of electrical raceway have been to facilitate the insertion and extraction of the conductors and to protect the conductors from mechanical injury (see Figure 15–18).

Today's raceway systems combine these functions with the need for a high degree of safety and greater distribution and utilization flexibility. It is the responsibility of the designer and installer to select the correct raceway for each application, and all things must be considered. *NEC®* Section 110–3 requires that during the judging of equipment, considerations such as the following shall be evaluated. The suitability of the installation must be considered and the use and conformity of equipment applicable with the provisions of the code; the suitability of the equipment for the specific purpose, environment and application of use; the suitability of the equipment as evidenced by listing or labeling. The mechanical strength and durability, including parts designed to enclose and protect the other equipment, is evaluated for suitability. The equipment is evaluated for adequate wire-bending and connection space, for electrical insulation, and heating effects under normal use and abnormal use. Equipment is evaluated for arcing effects. It is classified by type, size, voltage, current capacity, and specific use. Other factors include the practical safeguarding of persons likely to come in contact with the equipment. *NEC®* Section 110–3(b) states that listed and labeled equipment shall be used or installed in accordance with any instructions included in the listing or labeling. Therefore, improper selection of a raceway system or improper installation of that system is a *National Electrical Code®* violation, and the product may not provide the service or safety for which it is intended. Selection of a wiring method should include consideration of the type and design of the building and the number and capabilities of the occupants relative to escape in the event of a fire. Combustibility aspects of a raceway or other wiring method becomes important because the wiring system is installed throughout the building (see Figure 15–19). Also in some installations there are large quantities of wiring overhead. Often many branch circuits and feeders are routed from the equipment room, over or adjacent to corridors.

Figure 15–18 Raceway systems installed today are similar to those installed over fifty years ago. Many old systems are still being used today and continue to provide excellent physical protection. (Courtesy of Allied Tube & Conduit)

TYPES OF METALLIC WIRING METHODS SYSTEMS

Rigid Metal Conduit (*NEC®* Article 346)

NEC® Article 346 covers rigid metal conduit, also called galvanized rigid conduit (GRC), rigid metal conduit (RMC) and aluminum metal conduit (ARC) (see Figure 15–20). The most common is the heaviest weight, tubular steel conduit (GRC). It is smooth-walled and galvanized on both the inside (ID) and outside (OD). The galvanizing may be applied by the hot-dip process (dipping into a pot of molten zinc) or the electro-galvanizing process (electro-deposited zinc). Galvanizing provides protection against corrosion. The degree of corrosion resistance is influenced by environmental conditions such as moisture and chemical exposure.

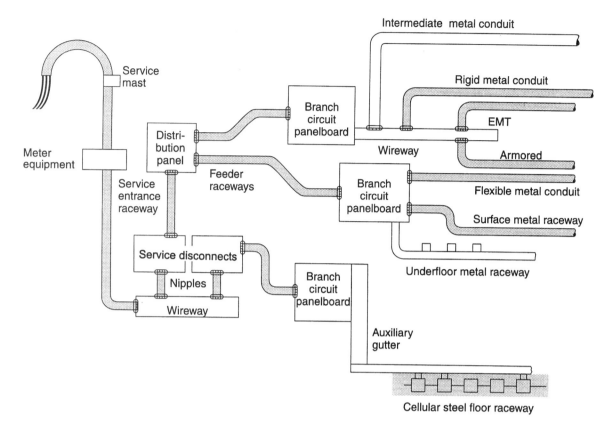

Figure 15–19 Wiring methods for 600 volts or less as per *NEC*® Chapter 3 and for over 600 volts as per Article 710.

Figure 15–20 Rigid steel conduit. (Courtesy of Allied Tube & Conduit)

Also covered under this article is nonferrous metal conduit. Both products are evaluated to the UL 6 Standard. The nonferrous product most commonly produced is aluminum rigid metal conduit and silicone bronze conduit and is evaluated to ANSI C80.5.

UL 6 and ANSI C80.1 are the standards for GRC and RMC. The international standard is IEC 981. These IEC soft metric conversions are proposed as the metric designations if or when the United States converts to the metric system. An example would be "½-inch GRC" would then become "16 GRC." These conduit types are currently listed and produced to the UL and ANSI standards. Both have been adopted by the federal government. The federal specification was formerly WW–C–581, Class 1/Type A. The dimensions and weights are shown in A–10 of the Appendix.

Galvanized rigid steel conduit is a threaded product, with threads conforming to ANSI B2.1. and *NEC*® Section 346–7(b). The UL listing requires that a listed piece of conduit have a coupling attached to one end. This coupling is electro-galvanized to avoid zinc buildup on the interior threads which might occur with hot-dipping. All rigid conduit is shipped with the coupling on one end and with the open threads on the opposite end protected by color-coded plastic caps. These caps are color-coded to provide easier field identification. The red caps indicate quarter sizes such as ¾-inch, 1¼-inch, and so on. The black caps are for ½-inch sizes such as ½-inch, 1½-inch, and so forth. The blue caps are for the even sizes such as 1, 2, 3, 4, 5, and 6 inches.

GRC is produced in trade sizes ½, ¾, 1, 1¼, 1½, 2, 2½, 3, 3½, 4, 5, and 6 inches. It should be noted that these trade designations are *not* actual OD or ID dimensions. The ID for ½-inch GRC, for example, is 0.632 inch; the OD for ½-inch is 0.840 inch. Neither dimension is the .500 one might expect. The actual diameters of all the recognized raceways can be found in *NEC®* Chapter 9 Table 4.

Article 346, and the *NEC®* in general, recognize the use of rigid steel conduit in *all* applications, including hazardous (classified) locations. The use of galvanized rigid steel conduit is only exempted where severe corrosive conditions exist. Even then, additional supplementary coatings, such as PVC-coated rigid metal conduit, can be used for those applications. Rigid steel conduit is the only conduit which can be used in all locations in accordance with the *NEC®* without exceptions. Steel rigid conduit and electrical metallic tubing are the most widely used raceway products in production today. It is common to find GRC conduit still in use in buildings built around the turn of the century and still providing good, safe service for the installation. Many buildings being remodeled today utilize the existing rigid metal conduit systems in those buildings for all or a portion of the new wiring being installed.

Galvanized steel rigid conduit evolved from black gas pipe which was already in place and a logical means of protecting those "newfangled" electrical conductors Tom Edison was touting! GRC is considered to be the premier electrical wiring method. It provides superior physical protection for conductors, is not affected by hot or cold temperatures (there are no code or standards limitations), and can be used indoors, outdoors, and underground, concealed or exposed.

Rigid metal conduit is a code-recognized, equipment-grounding conductor in accordance with *NEC®* Section 250–91(b).

Intermediate Metal Conduit (IMC)

Intermediate metal conduit, commonly known as IMC, is covered in *NEC®* Article 345. It was developed

Figure 15–21 IMC intermediate steel conduit. (Courtesy of Allied Tube & Conduit)

in the early 1970s by Allied Tube and Conduit. It originally was intended to be an entirely separate system, but has evolved with threads and dimensions which make it interchangeable with GRC (see Figure 15–21). In fact, the same couplings are used on both IMC and GRC. The *NEC®* Articles 345 and 346 are virtually identical, and the *NEC®* recognizes both products for exactly the same applications, without limitations.

The standards for listing and production of IMC are UL 1242 and ANSI C80.6. UL 6 has been adopted as a federal specification. IMC was formerly covered by WW–C–581 as Class 2/Type A (ANSI C80.6 has not yet completed the adoption process).

IMC is galvanized on the outside diameter (OD) and has a UL-approved corrosion resistant organic or inorganic coating on the internal diameter (ID). Although the IMC wall is less than the wall thickness of GRC, its method of manufacture and steel chemistry provide physical strength equivalent to, or often greater than, that of GRC, thus providing equivalent physical protection. Listed IMC is also a conduit threaded to ANSI B2.1 with an attached coupling. The color coding for the protective end caps is ¼-inch sizes green, ½-inch sizes yellow, and even sizes orange. This provides easy visual differentiation between IMC and GRC. Further identification is provided by marking the letters IMC on the pipe at regular intervals. The ID of IMC is slightly larger than that of GRC with ½-inch having a nominal .660 ID; the OD nominal for ½-inch is 0.815 inch, making wire pulling in IMC easier. The dimensions for nominal trade sizes of IMC can be found in Figure A–10 of the Appendix.

Wall thickness comparison is ½-inch IMC 0.070 inch and ½-inch GRC 0.104 inch. Compatible threads are possible because less steel is cut away in threading IMC. As with rigid conduit, each length is designed to be 10 feet, including the coupling.

IMC has become a very popular replacement for GRC in that it provides the equivalent physical protection, has a larger ID, and may be installed more easily because of its lighter weight. It is a totally different product than GRC and requires adjustment to common standard threading machines before threading and special bending equipment or adapters on standard bending tools. However, once a contractor is set up to install IMC, it usually is found to be easy to install.

Electrical Metallic Tubing (Article 348)

Electrical metallic tubing, commonly known as EMT, is the lightest weight tubular steel raceway manufactured. It is an unthreaded plain end product, which is joined together by the use of set-screw, indentation, or compression-type connectors and couplings. The coupling is not provided as a part of the listed EMT. EMT was developed before World War II, but did not gain popularity until the start of the war, when the entire nation was trying to conserve steel wherever possible. Soon after the use of EMT began and the tradesmen learned how to bend and install the product, it gained the massive popularity it enjoys today (see Figure 15–22). It is widely used throughout the industry for branch circuit and feeder raceways. It is very versatile in that it can be altered, reused, and redirected with ease because of the unthreaded design. The conductors can be inserted and extracted easily because of the very smooth interior, making conductor installation fast and easy. The internal diameter is sufficiently large to provide ease in pulling the number of conductors permitted in the *NEC*®. EMT has one of the larger IDs among the many raceway types being manufactured today. Despite the light-weight steel construction, it provides substantial physical pro-

Figure 15–22 EMT electrical metallic tubing. (Courtesy of Allied Tube & Conduit)

tection and may be used in most exposed locations, except where subject to *severe* physical damage.

The listing and production standards for EMT are UL 797 and ANSI C80.3. These standards have also been adopted by the federal government, replacing WW–C–563.

Electrical metallic tubing has a galvanized OD and a UL-approved corrosion-resistant organic or inorganic coating on the ID. For identification purposes, each ten-foot length is marked "EMT."

Again, the dimensions are trade size designations only, not actual ½-inch. ID is .620 inch and OD is a nominal 0.706 inch. Other weights and dimensions for electrical metallic tubing can be found in Figure A–10 of the Appendix.

The *NEC*® recognizes EMT for both exposed and concealed work. It may be used in most applications, except where subject to severe physical damage, or where subject to cinder concrete or cinder fill where subject to permanent moisture, unless protected on all sides by a layer of noncinder concrete at least 2 inches thick.

It is also limited in some hazardous locations. EMT is permitted to be installed in wet locations when used with proper fittings. Both rain-tight and concrete-tight fittings are available and must be used in wet locations and in poured concrete.

The support requirements for electrical metallic tubing can be found in *NEC®* Section 348–12. Bending shall be in accordance with Section 346–10 and shall not exceed four quarter bends or 360° total between pull points. In most instances, EMT shall be supported at least every 10 feet and within 3 feet of outlet boxes, junction boxes, and conduit bodies. In ceiling joists, bar joists, and other areas where structural members do not readily permit support within 3 feet, a distance of 5 feet from each outlet box, junction box, or conduit body shall be permitted.

SPECIAL FEATURE TUBULAR STEEL CONDUITS

Integral Coupling Steel Raceways

IMC and GRC may be purchased with an integral coupling, such as Kwik-Couple, which allows joining of two lengths of rigid or IMC together by turning the coupling rather than the conduit. This coupling performs similar to that of a union (Erickson) in that it can be tightened without rotating the conduit (see Figures 15–23 and 15–24). This type conduit can provide substantial savings where routing would cause difficulty in joining the conduit lengths together or several unions would be required due to the architectural design or obstructions in existing installations. Many installers integrate these special feature raceways with the traditional raceways to accomplish an easier and better installation. Where field bends or architectural appurtenances have made joining difficult or impossible without a union (Erickson) in the past, these new state-of-the-art products make installations easy and fast.

EMT is available with a belled-end fitting with installed set-screws or integral compression coupling which eliminates the need for a separate coupling. Substantial savings can be achieved in installed cost.

Coated Rigid Conduit & Intermediate Metal Conduit (IMC)

The coated metallic raceways that are manufactured today are manufactured in accordance with NEMA Standard RN–1 for "Polyvinyl Chloride (PVC) Externally Coated Galvanized Rigid Steel Conduit and Intermediate Metal Conduit." Coated metallic conduit must be installed in accordance with the article which covers the metallic conduit material used, such as galvanized rigid metal conduit must be installed as per *NEC®* Article 346 (see Figure 15–25). Intermediate metal conduit must be installed in accordance with *NEC®* Article 345. A copy of NEMA Standard RN–1 is available through the National Electrical Manufacturers Association, 2101 L St., NW, Washington, D.C. 20037. This product adopts in whole or in part ANSI C80.1 and C80.6. It also adopts ASTM D159/87 test methods for dielectric breakdown voltage and dielectric strain of solid electrical insulating materials at commercial power frequencies, D638–87, the test method for tensile properties of plastics; D1790–83, the test method for brittleness temperature of plastic film by impact; D2240–86, test method for rubber property durometer hardness; G6–83, test method for abrasion resistance of 5–ply coatings; G10–83, test method for specific bendability of pipeline coatings; G23–81, practice for operating light and water exposure apparatus (carbon-arc type) for exposure of nonmetallic materials; Underwriters Laboratories UL6; and UL1242.

PVC-coated rigid conduit was first introduced in 1961 by Robroy Industries. Today there are three major manufacturers of PVC-coated steel conduit. They are

Figure 15–23 Integral coupling steel rigid and IMC conduit meets Section 250–91(b) for grounding requirements of the *NEC®*. (Courtesy of Allied Tube & Conduit and Triangle Wire and Cable Company)

Galvanized Electrical Metallic Tubing (EMT)

Industry Standards:
UL 514B - Fittings for conduit and outlet boxes
UL 797 - Electrical metallic tubing
Federal Spec. WW-C-536-A - Conduit, metal, rigid electrical, thin-wall steel type
ANSI standard C80-3 - Electrical, metallic tubing zinc coated

Galvanized Electrical Metallic Tubing (EMT) Compression Uni-Couple

Industry Standards:
UL 514B - Fittings for conduit and outlet boxes
UL 797 - Electrical metallic tubing
Federal Spec. WW-C-536-A - Conduit, metal, rigid electrical, thin-wall steel type
ANSI standard C80-3 - Electrical, metallic tubing zinc coated

Figure 15–24 An example of two different types of integral coupling EMT electrical metallic tubing. (Courtesy of Allied Tube & Conduit and Triangle Wire and Cable Company)

Robroy Industries, Occidental Coating Company, and Perma-Cote Industries. These major manufacturers of coated conduit have recognized the need for a wiring method that provides complete corrosion protection and excellent installation qualities. PVC-coated conduit combines the strength of metal with the corrosion resistant qualities of a bonded plastic coating for a permanent, trouble free installation. It is an attractive product and identifies your electrical conduit system. In service it continues to create economies through complete and lasting corrosion protection. The adhesive qualities of the PVC coating effectively prevent corrosive fume seepage—there is no undercreep or corrosion travel. Overlapping pressure sealing sleeves on couplings and condulets create tight pressure sealed joints. You install it and forget it. There is no maintenance. The manufacturer begins by closely working with the designer and the installer to provide a complete material list. The coating process begins with listed rigid metal or intermediate metal conduit. Lighting fixtures can also be coated. Conduit bodies of all types and sizes are coated with overlapping pressure seals. Support channel, all hardware, straps, supports, etc., are coated and corrosion-proofed the same as the conduit. Motor-starting switches, unions, etc., can all be coated to provide a complete corrosion-resistant system. In addition, available from the manufacturer is a coating touch-up kit which consists of spray-on or paint-on protection that effectively repairs all damage to the original integrity and may even be of a heavier consistency than the original application (see Figure 15–26).

Figure 15–25 PVC-coated rigid conduit is ideal for high corrosive areas; areas subject to physical damage can be coated both inside and out. (Courtesy of Robroy Industries)

Figure 15–26 If PVC-coated conduit is damaged during installation, scrapes and cuts are easily patched with touch-up compounds available in spray cans, dabber, and brush on. (Courtesy of Occidental Coating Company)

This product has been widely accepted for industrial facilities, both aboveground exposed and for direct burial in all environments. There is no equal for this state-of-the-art raceway system. Coated conduit offers the combination of excellent physical protection by using UL-listed galvanized rigid steel conduit or galvanized intermediate steel conduit plus unequaled corrosion protection by the addition of a 40-mil thick PVC exterior coating. For severe environments, where corrosive vapors may also attack the conduit from the inside, a two-mil thick urethane interior coating can also be provided. These and other coatings can also be applied to rigid aluminum conduit to fit the requirements of each unique installations. **Note:** See Appendix for chemical resistance charts.

Liquidtight Flexible Metal Conduit

NEC® Article 351(a) and UL 360 liquidtight flexible metal conduit is manufactured in trade sizes ⅜-inch through 4-inch for conductors in circuits of 600 volts nominal or less. Liquidtight flexible metal conduit which is suitable for direct burial and poured in concrete is marked "Direct Burial." It is a raceway of circular cross section having an outer liquidtight non-metallic sunlight-resistant jacket over an inner flexible metal core with associated couplings, connectors, and fittings and approved for the installation of electrical conductors. Installations of liquidtight flexible metal conduit must comply with *NEC®* Article 300, 351, and specific sections where applicable of *NEC®* Articles 351, 501, 502, 503, and 553. The marking requirements for this product must comply with *NEC®* Section 110–21. It may be used where listed and marked for the purpose for direct burial in earth and exposed and concealed work where the conditions of operation and maintenance require flexibility or protection from liquids, vapors, or solids. It shall not be used where subject to physical damage, or where any combination of ambient and/or conductor temperature will produce an operating temperature in excess of that for which the material is approved. Where it is installed as a fixed raceway, it shall be secured at intervals not exceeding 4½ feet and within 12 inches on each side of every outlet, box, junction box, cabinet or fitting, except where permitted to be fished or not exceeding 3 feet in length for flexibility, and not exceeding 6 feet in length for fixture tap conductors. It is permitted as a grounding conductor where both the conduit and the fittings are approved for grounding. Where a bonding jumper is required, it must be installed in accordance with *NEC®* Section 250–79. Where liquidtight metal conduit is used for flexibility, an equipment-grounding conductor must be installed. It is only permitted as a grounding means in 1¼-inches trade sizes and smaller if the total length of the liquidtight flexible metal conduit and any ground return path is 6 feet or less and the conduit is terminated in fittings listed for grounding and the circuit conductors contained therein are protected by an overcurrent device rated at 20 amperes or less for ⅜-inch and ½-inch trade sizes, and 60 amperes or less for ¾-inch through 1¼-inches trade sizes. Where used as a fixed wiring method, there shall not be more than the equivalent of four quarter bends, 360° total, between pull points, e.g., conduit bodies and boxes. Angle conductors are not permitted for use in concealed raceway installations.

Rigid Nonmetallic Conduit Schedule 40 and Schedule 80

Rigid nonmetallic conduit (PVC) (Schedule 40 and Schedule 80), intended for installation in accordance with *NEC®* Article 347 and UL 651, was originally introduced as an underground duct for use by electrical utilities in the late 1950s. At that time the duct was produced from high-impact polystyrene, which had excellent physical properties and corrosion resistance (see Figure 15–27). However, the styrene material did not have good fire resistant properties, and by the early 1960s polyvinyl chloride (PVC) became the preferred material. UL first listed PVC Schedule 40 rigid nonmetallic conduit in 1962. The *NEC®* first recognized its use only for underground installations in 1968. In 1971 the *NEC®* expanded approved uses to include aboveground and in-building applications. Article 347 as it appears in the 1996 *NEC®* remains virtually unchanged from the original 1971 edition (see Figure 15–28).

Figure 15–27 Underground nonmetallic duct bank installation. (Courtesy of Carlon, a Lampson Sessions Company)

Rigid nonmetallic conduit, fittings, and accessories also referred to as RNMC are manufactured to NEMA TC–2 Federal Specifications WC1094A and UL 651 specifications and carry respective UL listings and UL labels. RNMC is nonconductive, sunlight-resistant, and UL-listed for exposed or outdoor usage (the use of expansion fittings allows the system to expand and contract with temperature variations), and it will not rust or corrode. The most commonly used RNMC is Schedule 40 rigid heavy wall PVC for underground applications encased in concrete or direct burial, exposed and concealed, rated for use with 90°C-conductors, and is manufactured in nominal sizes ½-inch through 6-inches. It is produced in 10-foot lengths, but may be produced in shorter or longer than ten-foot lengths, with or without belled ends. Where subject to physical damage for aboveground or belowground applications, RNMC Schedule 80 extra heavy wall PVC-80 **must be used.** PVC Schedule 80 is also produced in ½-inch through 6-inch-sizes, with or without belled ends. *NEC®* Article 347 permits rigid nonmetallic conduit and fittings to be used under the following conditions, an informational fine print note advises that extreme cold may cause some nonmetallic conduits to become brittle and, therefore, more susceptible to damage from physical contact.

For underground applications encased in concrete or direct burial. Also for use in exposed or concealed applications aboveground.
- UL Listed
- Sunlight resistant
- Rated for use with 90°C conductors
- Superior weathering characteristics

For use in aboveground and belowground applications that are subject to physical damage.
- UL Listed
- Sunlight resistant
- Rated for use with 90°C conductors
- Superior weathering characteristics

Figure 15–28 Installation as per *NEC®* Article 347 rigid nonmetallic conduit. (Courtesy of Carlon, a Lampson Sessions Company)

Uses Permitted: (Varies by UL listing. Some nonmetallic conduits are listed and marked for underground use only.)

RNMC is permitted to be installed concealed in walls, floors, and ceilings in locations subject to severe corrosive influences, as covered by *NEC®* Section 300–6, or where subject to chemicals for which the materials are specifically approved. For chemicals for which PVC is not acceptable, the designer or installer should select an appropriate wiring method from Chapter 3 of the *National Electrical Code®*. **Note:** see Figure A–13 in the Appendix. The installations must comply with the specific *NEC®* article and, in addition, be installed in accordance with the specific listing instructions as applicable.

In wet locations, such as dairies, laundries, canneries, and other wet locations or where walls are frequently washed, the entire conduit system shall be so installed and equipped as to prevent water from entering the system. All supports, bolts, straps, screws, etc., shall be of a corrosion-resistant material or protected against corrosion by corrosion-resistant materials. RNMC can be used in dry and damp locations and exposed where not subject to physical damage, except Schedule 80 may be installed in locations subject to physical damage. Rigid nonmetallic conduit may be installed underground as per *NEC®* Sections 300–5 and 710–3(b). **Note:** (Schedule 80 has less interior space and a reduction in wire fill is necessary). See *NEC®* Chapter 9 Tables 3A through 3F. Underground installations are to be made, as per *NEC®* Sections 300–5 and 710–3(b).

Rigid nonmetallic conduit is not permitted to be used in hazardous locations except as per *NEC®* Sections 504–20, 515–8 and 515–5, and Class I Division 2 locations as permitted in the Exception to Section 501–4(b). It is not permitted for the support of fixtures or other equipment unless identified specifically for such use. It is not permitted where subject to ambient temperatures exceeding 50°C (122°F). It is not permitted in theaters or similar locations, except under limited conditions as provided in *NEC®* Articles 518 and 520. Rigid nonmetallic conduit is not permitted to be used in environmental air spaces as detailed in *NEC®* Article 300–22.

Rigid nonmetallic conduit products are joined together by means of solvent cement. Sizes ½-inch through 1½-inches should be cut square using a fine-toothed saw and deburred. For sizes 2 inches through 6 inches, a miter box or similar saw guide should be utilized to keep the material steady. After cutting and deburring, wipe the ends clean of dust, dirt, and shavings. The joining process is as follows: Be sure that the conduit end is clean and dry. Place a coating of primer/cleaner on the end and then on its mating part with a dauber. Thoroughly coat the surfaces to be mated. Allow primer/cleaner a few seconds to soften the PVC surface (the time may need to be adjusted, depending upon the temperature). Apply a coat of solvent cement to the end of the conduit for the length of the socket to be attached. Push conduit firmly into fitting while rotating conduit slightly about one-quarter turn to spread cement evenly. Allow the joint to set approximately 10 minutes. Most manufacturers recommend specific solvent cement. The cement is prepared for their products, compounds, and tolerances, and substitutions should not be made as adverse effects can occur. In situations requiring extremely fast setting, or in low temperature or difficult installation conditions, all-weather quick-set cement is available and should be used. Rigid nonmetallic conduit may be bent in the field. *NEC®* Sections 347–13 and 347–15 specify how the bends are to be made and the number permitted in a run. They must be made so that the conduit is not damaged and the internal diameter of the conduit will not effectively be reduced. It shall only be made with bending equipment identified for the purpose. For more information on bending, see Chapter 3. The number of conductors permitted shall not exceed that permitted by the percentage fill in *NEC®* Table 1, Chapter 9.

> Rigid nonmetallic conduit, in accordance with Section 347–13, shall only be bent using bending equipment identified for the purpose.

REVIEW QUESTIONS

(Answers to some of these questions may be found in the introduction, previous chapters, and 1996 *NEC®*.)

1. Is an electrical equipment room containing switchboards and panelboards required to be illuminated?
   Answer: Reference:

2. The working space in front of enclosed electrical equipment operating at 480 volts must have a *width,* when facing the equipment, of at least _____ inches.
   Answer: Reference:

3. A conduit nipple of 18 inches in length is installed between two boxes. This nipple may be filled to a maximum of _____ percent without derating the conductors.
   Answer: Reference:

4. Circuit conductors run in electrical nonmetallic tubing (ENT) may not exceed _____ volts.
   Answer: Reference:

5. When a metal raceway is used as physical protection for a grounding electrode conductor, must this metallic raceway be bonded, even though the grounding electrode conductor is bare copper? If the answer is yes, how must this metal raceway be bonded?
   Answer: Reference:

6. You have run into a problem in making a new service installation and need to know the requirements of the meter enclosure on the outside of the building. Which article of the *National Electrical Code®* covers meter enclosures?
   Answer: Reference:

7. In making an installation of over 200 feet, a concern is voltage drop. Where does one obtain the resistance for conductors to be installed in a raceway or in direct burial for the purposes of calculating voltage drop?
   Answer: Reference:

8. If the clearances that are required by Section 110–16 can be maintained in front of a panelboard or a service panel to be installed in a bathroom in an industrial plant office building, would this installation be permitted by the *NEC®*?
   Answer: Reference:

9. The Code generally prohibits splicing a grounding electrode conductor. Is this conductor permitted to be tapped? If so, how are these taps sized?
   Answer: Reference:

10. What is a separately derived system? Where is it mentioned in the Code?
    Answer: Reference:

11. What is the maximum permitted contraction or expansion for a run of Schedule 40 PVC conduit before the installation of an expansion coupling is required?
    Answer: Reference:

12. What is the maximum size flexible metallic tubing manufactured today?
    Answer: Reference:

13. I have just encountered a nonmetallic wireway to be installed in a wet location. Which section of the *National Electrical Code®* covers this nonmetallic wireway?
    Answer: Reference:

14. A motor is to be installed in a Class I Division 2 location. It is to be connected with 3 feet of ¾-inch liquidtight flexible metal conduit. Is it permissible to use the liquidtight flexible metal conduit as the grounding path for this installation?
    Answer:					Reference:

15. A rigid metal conduit contains three circuits: two 150 amp, 3-phase circuits, and one 300 amp, single-phase circuit. The load side service equipment bonding jumper for this conduit must be at least size _____ AWG copper.
    Answer:					Reference:

16. Are EMT set-screw connectors approved for grounding of service raceways, or are jumpers required to bond around them?
    Answer:					Reference:

17. Does the Code require interior gas piping systems to be bonded to the grounding electrode system?
    Answer:					Reference:

18. In a run of conduit supplying a motor, it is necessary to install 4 feet of flexible metal conduit where the raceway leaves the panel to get around a coil. When the conduit reaches the motor, an additional 3 feet of flexible metal conduit is employed for convenience and flexibility. Is an equipment-grounding conductor required in this metal raceway?
    Answer:					Reference:

Glossary

ANSI: American National Standards Institute; ANSI is an umbrella organization for other standards writing organizations. After a standard has been developed and gone through the consensus process, it is then processed as an American National Standard.

Authority having jurisdiction (AHJ): The person or discipline responsible for enforcement of the Code and having the responsibility for making interpretations of the rules, for deciding upon the approval of equipment and materials, and for granting the special permission contemplated in a number of the rules.

API: the American Petroleum Institute; they have unique recommendations for the classification of areas dealing with petroleum facilities, such as API IRP 500–1991.

ASTM E119: Standard Method of Fire Test for Building Construction and Materials.

BOCA: Building Officials and Code Administrators International, Country Club Hills, IL; the national codes are adopted throughout the eastern and upper mid-west. They produce the National Building Code, Plumbing, Mechanical, Fire, etc.

Consensus process: All changes are made in a democratic fashion based on public input and voted and acted on by a balanced committee of voluntary experts or representatives from all facets of the standard.

Fine Print Notes (FPN): Explanatory information and references are in the form of Fine Print Notes which are identified (FPN). These FPNs are not enforceable.

Green Book: The UL *Electrical Constructions Material Directory.*

Flash point: Flash point of a liquid shall mean the minimum temperature at which it gives off vapor in sufficient concentration to form an ignitable mixture of air near the surface of the liquid within the vessel as specified by an appropriate test procedure and apparatus.

Example: The flash point of a liquid having a viscosity less than 45 SUS at 100°F (37.8°C) and a flash point below 200°F (93.4°C) shall be determined in accordance with ASTM D56, Standard Method of Test for Flash Point by the TAG Closed Tester. (See definition of flash point in NFPA 30A or other NFPA standards for flammable materials.)

GRC: Rigid metal conduit as covered in *NEC®* Article 346 galvanized steel conduit.

Hazardous atmosphere: atmospheres containing concentrations of flammable vapors, gases or liquids, or combustible dusts, or ignitable fibers or flyings.

ICBO: International Conference of Building Officials, Whittier, CA. The uniform codes are adopted throughout the western and mid-western states. They produce all of the uniform codes, such as the building, mechanical, plumbing, fire.

ID: Internal diameter.

IEC: International Electrotechnical Commission, based in Europe; writes standards for international use. Often thought of as a European organization, however, the United States actively participates in this standards writing body, along with representatives from many other countries.

Listing mark: Often called the "bug" or trademark, generally the listing agency's initials in a unique configuration.

Label: The information label on the product, usually containing the listing mark.

NEMA: The National Electrical Manufacturers Association located in Rosslyn, VA.

National Electrical Code® (NEC®): The most widely used electrical code in the world is the *NEC®*. The official designation for the *National Electrical Code®* is ANSI/NFPA 70.

NFPA: National Fire Protection Association; a standards writing organization located in Quincy, MA, which is responsible for many fire and safety standards.

Nonincendive circuits: A circuit in which any arc or thermal effect produced, under intended operating conditions of the equipment or due to the opening, shorting, or grounding of the field wiring, is not capable, under specified test conditions, of igniting the flammable gas, vapor, or dust-air mixture.

NRTL: Nationally Recognized Testing Laboratory. These are laboratories that are recognized by OSHA as competent testing laboratories.

OD: Outside diameter.

OSHA: Occupational Safety and Health Administration (Act), intended to protect employees from hazards in the workplace.

Red Book: *Hazardous Location Equipment* will provide valuable information which can help avoid the misapplication of materials.

RMC: Rigid metal conduit as covered in *NEC®* Article 346; usually galvanized steel or aluminum, but may be other metals, such as stainless steel or brass.

SBCCI: Southern Building Code Congress International Inc., Birmingham, AL. The Standard Codes are adopted throughout the southern states. They produce all of the Standard Codes, Building Code, Mechanical Code, Plumbing, Fire, etc.

White Book: General Information for *Electrical Construction, Hazardous Location, and Electric Heating and Air-Conditioning Equipment.*

Appendix

Hazardous Substances Tables

A–1 Hazardous Substances Flammable Gases and Vapors

A–2 Hazardous Substances Combustible Dusts

A–3 Recommended Uses for Raceways and Other Wiring Methods

A–4 Galvanized Rigid Conduit Specifications

A–5 Intermediate Metal Conduit (IMC) Specifications

A–6 Electrical Metallic Tubing (EMT) Specifications

A–7 Rigid Nonmetallic Conduit (Schedule 40) Specifications

A–8 Rigid Nonmetallic Conduit (Schedule 80) Specifications

A–9 Approximate Diameters for Standard Conduit Locknuts and Bushings

A–10 Recommended Spacings for Conduit and Tubing

A–11 Basic Electrical AC and DC Formulas

A–12 Electrical Symbols

A–13 Chemical Resistance Chart for PVC Conduit & Tubing

A–14 Chemical Resistance Chart for Urethane Coatings

A–15 Chemical Resistance Chart for PVC Coatings

Class 1* Group	Substance	Auto-Ignition Temp. °F	Auto-Ignition Temp. °C	Flash Point °F	Flash Point °C	Explosive Limits** Per Cent by Volume Lower	Explosive Limits** Per Cent by Volume Upper	Vapor** Density (Air Equals 1.0)
C	Acetaldehyde	347	175	-38	-39	4.0	60	1.5
D	Acetic Acid	867	464	103	39	4.0	19.9 @ 200° F	2.1
D	Acetic Anhydride	600	316	120	49	2.7	10.3	3.5
D	Acetone	869	465	-4	-20	2.5	13	2.0
D	Acetone Cyanohydrin	1270	688	165	74	2.2	12.0	2.9
D	Acetonitrile	975	524	42	6	3.0	16.0	1.4
A	Acetylene	581	305	gas	gas	2.5	100	0.9
B(C)	Acrolein (inhibited)[1]	455	235	-15	-26	2.8	31.0	1.9
D	Acrylic Acid	820	438	122	50	2.4	8.0	2.5
D	Acrylonitrile	898	481	32	0	3.0	17	1.8
D	Adiponitrile	—	—	200	93	—	—	—
C	Allyl Alcohol	713	378	70	21	2.5	18.0	2.0
D	Allyl Chloride	905	485	-25	-32	2.9	11.1	2.6
B(C)	Allyl Glycidyl Ether[1]	—	—	—	—	—	—	—
D	Ammonia[2]	928	498	gas	gas	15	28	0.6
D	n-Amyl Acetate	680	360	60	16	1.1	7.5	4.5
D	sec-Amyl Acetate	—	—	89	32	—	—	4.5
D	Aniline	1139	615	158	70	1.3	11	3.2
D	Benzene	928	498	12	-11	1.3	7.9	2.8
D	Benzyl Chloride	1085	585	153	67	1.1	—	4.4
B(D)	1,3-Butadiene[1]	788	420	gas	gas	2.0	12.0	1.9
D	Butane	550	288	gas	gas	1.6	8.4	2.0
D	1-Butanol	650	343	98	37	1.4	11.2	2.6
D	2-Butanol	761	405	75	24	1.7 @ 212° F	9.8 @ 212° F	2.6
D	n-Butyl Acetate	790	421	72	22	1.7	7.6	4.0
D	iso-Butyl Acetate	790	421	—	—	—	—	—
D	sec-Butyl Acetate	—	—	88	31	1.7	9.8	4.0
D	t-Butyl Acetate	—	—	—	—	—	—	—
D	n-Butyl Acrylate (inhibited)	559	293	118	48	1.5	9.9	4.4
C	n-Butyl Formal	—	—	—	—	—	—	—
B(C)	n-Butyl Glycidyl Ether[1]	—	—	—	—	—	—	—
C	Butyl Mercaptan	—	—	35	2	—	—	3.1
D	t-Butyl Toluene	—	—	—	—	—	—	—
D	Butylamine	594	312	10	-12	1.7	9.8	2.5
D	Butylene	725	385	gas	gas	1.6	10.0	1.9
C	n-Butyraldehyde	425	218	-8	-22	1.9	12.5	2.5
D	n-Butyric Acid	830	443	161	72	2.0	10.0	3.0
—[3]	Carbon Disulfide	194	90	-22	-30	1.3	50.0	2.6
C	Carbon Monoxide	1128	609	gas	gas	12.5	74.0	1.0
C	Chloroacetaldehyde	—	—	—	—	—	—	—
D	Chlorobenzene	1099	593	82	28	1.3	9.6	3.9
C	1-Chloro-1-Nitropropane	—	—	144	62	—	—	4.3
D	Chloroprene	—	—	-4	-20	4.0	20.0	3.0
D	Cresol	1038-1110	559-599	178-187	81-86	1.1-1.4	—	—
C	Crotonaldehyde	450	232	55	13	2.1	15.5	2.4
D	Cumene	795	424	96	36	0.9	6.5	4.1
D	Cyclohexane	473	245	-4	-20	1.3	8.0	2.9
D	Cyclohexanol	572	300	154	68	—	—	3.5
D	Cyclohexanone	473	245	111	44	1.1 @ 212° F	9.4	3.4
D	Cyclohexene	471	244	<20	<-7	—	—	2.8
D	Cyclopropane	938	503	gas	gas	2.4	10.4	1.5
D	p-Cymene	817	436	117	47	0.7 @ 212° F	5.6	4.6
C	n-Decaldehyde	—	—	—	—	—	—	—
D	n-Decanol	550	288	180	82	—	—	5.5
D	Decene	455	235	<131	<55	—	—	4.84
D	Diacetone Alcohol	1118	603	148	64	1.8	6.9	4.0
D	o-Dichlorobenzene	1198	647	151	66	2.2	9.2	5.1
D	1,1-Dichloroethane	820	438	22	-6	5.6	—	—
D	1,2-Dichloroethylene	860	460	36	2	5.6	12.8	3.4
C	1,1-Dichloro-1-Nitroethane	—	—	168	76	—	—	5.0
D	1,3-Dichloropropene	—	—	95	35	5.3	14.5	3.8
C	Dicyclopentadiene	937	503	90	32	—	—	—
D	Diethyl Benzene	743-842	395-450	133-135	56-57	—	—	4.6
C	Diethyl Ether	320	160	-49	-45	1.9	36.0	2.6
C	Diethylamine	594	312	-9	-23	1.8	10.1	2.5
C	Diethylaminoethanol	—	—	—	—	—	—	—
C	Diethylene Glycol Monobutyl Ether	442	228	172	78	0.85	24.6	5.6
C	Diethylene Glycol Monomethyl Ether	465	241	205	96	—	—	—
D	Di-isobutyl Ketone	745	396	120	49	0.8 @ 200° F	7.1 @ 200° F	4.9
D	Di-isobutylene	736	391	23	-5	0.8	4.8	3.9
C	Di-isopropylamine	600	316	30	-1	1.1	7.1	3.5
C	N-N-Dimethyl Aniline	700	371	145	63	—	—	4.2
D	Dimethyl Formamide	833	455	136	58	2.2 @ 212° F	15.2	2.5
D	Dimethyl Sulfate	370	188	182	83	—	—	4.4
C	Dimethylamine	752	400	gas	gas	2.8	14.4	1.6
C	1,4-Dioxane	356	180	54	12	2.0	22	3.0
D	Dipentene	458	237	113	45	0.7 @ 302° F	6.1 @ 302° F	4.7
C	Di-n-propylamine	570	299	63	17	—	—	3.5
C	Dipropylene Glycol Methyl Ether	—	—	185	85	—	—	5.11
D	Dodecene	491	255	—	—	—	—	—

Figure A-1(A) Properties of gases, liquids, and vapors. (Courtesy of Crouse-Hinds, Division of Cooper Industries)

Appendix 165

Class 1* Group	Substance	Auto-Ignition Temp. °F	Auto-Ignition Temp. °C	Flash Point °F	Flash Point °C	Explosive Limits** Per Cent by Volume Lower	Explosive Limits** Per Cent by Volume Upper	Vapor** Density (Air Equals 1.0)
C	Epichlorohydrin	772	411	88	31	3.8	21.0	3.2
D	Ethane	882	472	gas	gas	3.0	12.5	1.0
D	Ethanol	685	363	55	13	3.3	19	1.6
D	Ethyl Acetate	800	427	24	-4	2.0	11.5	3.0
D	Ethyl Acrylate (inhibited)	702	372	50	10	1.4	14	3.5
D	Ethyl sec-Amyl Ketone	—	—	—	—	—	—	—
D	Ethyl Benzene	810	432	70	21	1.0	6.7	3.7
D	Ethyl Butanol	—	—	—	—	—	—	—
D	Ethyl Butyl Ketone	—	—	115	46	—	—	4.0
D	Ethyl Chloride	966	519	-58	-50	3.8	15.4	2.2
D	Ethyl Formate	851	455	-4	-20	2.8	16.0	2.6
D	2-Ethyl Hexanol	448	231	164	73	0.88	9.7	4.5
D	2-Ethyl Hexyl Acrylate	485	252	180	82	—	—	—
C	Ethyl Mercaptan	572	300	<0	<-18	2.8	18.0	2.1
C	n-Ethyl Morpholine	—	—	—	—	—	—	—
C	2-Ethyl-3-Propyl Acrolein	—	—	155	68	—	—	4.4
D	Ethyl Silicate	—	—	125	52	—	—	7.2
D	Ethylamine	725	385	<0	<-18	3.5	14.0	1.6
C	Ethylene	842	450	gas	gas	2.7	36.0	1.0
D	Ethylene Chlorohydrin	797	425	140	60	4.9	15.9	2.8
D	Ethylene Dichloride	775	413	56	13	6.2	16	3.4
C	Ethylene Glycol Monobutyl Ether	460	238	143	62	1.1 @ 200° F	12.7 @ 275° F	4.1
C	Ethylene Glycol Monobutyl Ether Acetate	645	340	160	71	0.88 @ 200° F	8.54 @ 275° F	—
C	Ethylene Glycol Monoethyl Ether	455	235	110	43	1.7 @ 200° F	15.6 @ 200° F	3.0
C	Ethylene Glycol Monoethyl Ether Acetate	715	379	124	52	1.7	—	4.72
D	Ethylene Glycol Monomethyl ether	545	285	102	39	1.8 @ STP	14 @ STP	2.6
B(C)	Ethylene Oxide¹	804	429	-20	-28	3.0	100	1.5
D	Ethylenediamine	725	385	93	34	4.2	14.4	2.1
C	Ethylenimine	608	320	12	-11	3.6	46.0	1.5
C	2-Ethylhexaldehyde	375	191	112	44	0.85 @ 200° F	7.2 @ 275° F	4.4
B	Formaldehyde (Gas)	795	429	gas	gas	7.0	73	1.0
D	Formic Acid (90%)	813	434	122	50	18	57	—
D	Fuel Oils	410-765	210-407	100-336	38-169	0.7	5	—
C	Furfural	600	316	140	60	2.1	19.3	3.3
C	Furfuryl Alcohol	915	490	167	75	1.8	16.3	3.4
D	Gasoline	536-880	280-471	-36 to -50	-38 to -46	1.2-1.5	7.1-7.6	3-4
D	Heptane	399	204	25	-4	1.05	6.7	3.5
D	Heptene	500	260	<32	<0	—	—	3.39
D	Hexane	437	225	-7	-22	1.1	7.5	3.0
D	Hexanol	—	—	145	63	—	—	3.5
D	2-Hexanone	795	424	77	25	—	8	3.5
D	Hexenes	473	245	<20	<-7	—	—	3.0
D	sec-Hexyl Acetate	—	—	—	—	—	—	—
C	Hydrazine	74-518	23-270	100	38	2.9	9.8	1.1
B	Hydrogen	968	520	gas	gas	4.0	75	0.1
C	Hydrogen Cyanide	1000	538	0	-18	5.6	40.0	0.9
C	Hydrogen Selenide	—	—	—	—	—	—	—
C	Hydrogen Sulfide	500	260	gas	gas	4.0	44.0	1.2
D	Isoamyl Acetate	680	360	77	25	1.0 @ 212° F	7.5	4.5
D	Isoamyl Alcohol	662	350	109	43	1.2	9.0 @ 212° F	3.0
D	Isobutyl Acrylate	800	427	86	30	—	—	4.42
C	Isobutyraldehyde	385	196	-1	-18	1.6	10.6	2.5
C	Isodecaldehyde	—	—	185	85	—	—	5.4
D	Iso-octyl Alcohol	—	—	180	82	—	—	—
C	Iso-octyl Aldehyde	387	197	—	—	—	—	—
D	Isophorone	860	460	184	84	0.8	3.8	—
D	Isoprene	428	220	-65	-54	1.5	8.9	2.4
D	Isopropyl Acetate	860	460	35	2	1.8 @ 100° F	8	3.5
D	Isopropyl Ether	830	443	-18	-28	1.4	7.9	3.5
C	Isopropyl Glycidyl Ether	—	—	—	—	—	—	—
D	Isopropylamine	756	402	-35	-37	—	—	2.0
D	Kerosene	410	210	110-162	43-72	0.7	5	—
D	Liquefied Petroleum Gas	761-842	405-450	—	—	—	—	—
B	Manufactured Gas (containing more than 30% H₂ by volume)	—	—	—	—	—	—	—
D	Mesityl Oxide	652	344	87	31	1.4	7.2	3.4
D	Methane	999	630	gas	gas	5.0	15.0	0.6
D	Methanol	725	385	52	11	6.0	36	1.1
D	Methyl Acetate	850	454	14	-10	3.1	16	2.8
D	Methyl Acrylate	875	468	27	-3	2.8	25	3.0
D	Methyl Amyl Alcohol	—	—	106	41	1.0	5.5	—
D	Methyl n-Amyl Ketone	740	393	102	39	1.1 @ 151° F	7.9 @ 250° F	3.9
C	Methyl Ether	662	350	gas	gas	3.4	27.0	1.6
D	Methyl Ethyl Ketone	759	404	16	-9	1.7 @ 200° F	11.4 @ 200° F	2.5
D	2-Methyl-5-Ethyl Pyridine	—	—	155	68	1.1	6.6	4.2
C	Methyl Formal	460	238	—	—	—	—	—
D	Methyl Formate	840	449	-2	-19	4.5	23	2.1

Figure A–1(B) Properties of gases, liquids, and vapors. (Courtesy of Crouse-Hinds, Division of Cooper Industries)

Class 1* Group	Substance	Auto-Ignition Temp. °F	Auto-Ignition Temp. °C	Flash Point °F	Flash Point °C	Explosive Limits Per Cent by Volume Lower	Explosive Limits Per Cent by Volume Upper	Vapor Density (Air Equals 1.0)
D	Methyl Isobutyl Ketone	840	440	64	18	1.2 @ 200° F	8.0 @ 200° F	3.5
D	Methyl Isocyanate	994	534	19	-7	5.3	26	1.97
C	Methyl Mercaptan	—	—	—	—	3.9	21.8	1.7
D	Methyl Methacrylate	792	422	50	10	1.7	8.2	3.6
D	2-Methyl-1-Propanol	780	416	82	28	1.7 @ 123° F	10.6 @ 202° F	2.6
D	2-Methyl-2-Propanol	892	478	52	11	2.4	8.0	2.6
D	alpha-Methyl Styrene	1066	574	129	54	1.9	6.1	—
C	Methylacetylene	—	—	gas	gas	1.7	—	1.4
C	Methylacetylene-Propadiene (stabilized)	—	—	—	—	—	—	—
D	Methylamine	806	430	gas	gas	4.9	20.7	1.0
D	Methylcyclohexane	482	250	25	-4	1.2	6.7	3.4
D	Methylcyclohexanol	565	296	149	65	—	—	3.9
D	o-Methylcyclohexanone	—	—	118	48	—	—	3.9
D	Monoethanolamine	770	410	185	85	—	—	2.1
D	Monoisopropanolamine	705	374	171	77	—	—	2.6
C	Monomethyl Aniline	900	482	185	85	—	—	3.7
C	Monomethyl Hydrazine	382	194	17	-8	2.5	92	1.6
C	Morpholine	590	310	98	37	1.4	11.2	3.0
D	Naphtha (Coal Tar)	531	277	107	42	—	—	—
D	Naphtha (Petroleum)[4]	550	288	<0	<-18	1.1	5.9	2.5
D	Nitrobenzene	900	482	190	88	1.8 @ 200° F	—	4.3
C	Nitroethane	778	414	82	28	3.4	—	2.6
C	Nitromethane	785	418	95	35	7.3	—	2.1
C	1-Nitropropane	789	421	96	36	2.2	—	3.1
C	2-Nitropropane	802	428	75	24	2.6	11.0	3.1
D	Nonane	401	205	88	31	0.8	2.9	4.4
D	Nonene	—	—	78	26	—	—	4.35
D	Nonyl Alcohol	—	—	165	74	0.8 @ 212° F	6.1 @ 212° F	5.0
D	Octane	403	206	56	13	1.0	6.5	3.9
D	Octene	446	230	70	21	—	—	3.9
D	n-Octyl Alcohol	—	—	178	81	—	—	4.5
D	Pentane	470	243	<-40	<-40	1.5	7.8	2.5
D	1-Pentanol	572	300	91	33	1.2	10.0 @ 212° F	3.0
D	2-Pentanone	846	452	45	7	1.5	8.2	3.0
D	1-Pentene	527	275	0	-18	1.5	8.7	2.4
D	Phenylhydrazine	—	—	190	88	—	—	—
D	Propane	842	450	gas	gas	2.1	9.5	1.6
D	1-Propanol	775	413	74	23	2.2	13.7	2.1
D	2-Propanol	750	399	53	12	2.0	12.7 @ 200° F	2.1
D	Propiolactone	—	—	165	74	2.9	—	2.5
C	Propionaldehyde	405	207	-22	-30	2.6	17	2.0
D	Propionic Acid	870	466	126	52	2.9	12.1	2.5
D	Propionic Anhydride	545	285	145	63	1.3	9.5	4.5
D	n-Propyl Acetate	842	450	55	13	1.7 @ 100° F	8	3.5
C	n-Propyl Ether	419	215	70	21	1.3	7.0	3.53
B	Propyl Nitrate	347	175	68	20	2	100	—
D	Propylene	851	455	gas	gas	2.0	11.1	1.5
D	Propylene Dichloride	1035	557	60	16	3.4	14.5	3.9
B(C)	Propylene Oxide[1]	840	449	-35	-37	2.3	36	2.0
D	Pyridine	900	482	68	20	1.8	12.4	2.7
D	Styrene	914	490	88	31	1.1	7.0	3.6
C	Tetrahydrofuran	610	321	6	-14	2.0	11.8	2.5
D	Tetrahydronaphthalene	725	385	160	71	0.8 @ 212° F	5.0 @ 302° F	4.6
C	Tetramethyl Lead	—	—	100	38	—	—	6.5
D	Toluene	896	480	40	4	1.2	7.1	3.1
D	Tridecene	—	—	—	—	—	—	—
C	Triethylamine	480**	249**	16	-9	1.2	8.0	3.5
D	Triethylbenzene	—	—	181	83	—	—	5.6
D	Tripropylamine	—	—	105	41	—	—	4.9
D	Turpentine	488	253	95	35	0.8	—	—
D	Undecene	—	—	—	—	—	—	—
C	Unsymmetrical Dimethyl Hydrazine (UDMH)	480	249	5	-15	2	95	2.0
C	Valeraldehyde	432	222	54	12	—	—	3.0
D	Vinyl Acetate	756	402	18	-8	2.6	13.4	3.0
D	Vinyl Chloride	882	472	gas	gas	3.6	33.0	2.2
D	Vinyl Toluene	921	494	120	49	—	11.0	4.1
D	Vinylidene Chloride	1058	570	-19	-28	6.5	15.5	3.4
D	Xylenes	867-984	464-529	81-90	27-32	1.0-1.1	7.0	3.7

[1] If equipment is isolated by sealing all conduit ½ in. or larger, in accordance with Section 501-5(a) of NFPA 70, *National Electrical Code*, equipment for the group classification shown in parentheses is permitted.

[2] For classification of areas involving Ammonia, see *Safety Code for Mechanical Refrigeration*, ANSI/ASHRAE 15, and *Safety Requirements for the Storage and Handling of Anhydrous Ammonia*, ANSI/CGA G2.1.

[3] Certain chemicals may have characteristics that require safeguards beyond those required for any of the above groups. Carbon disulfide is one of these chemicals because of its low autoignition temperature and the small joint clearance to arrest its flame propagation.

[4] Petroleum Naphtha is a saturated hydrocarbon mixture whose boiling range is 20° to 135° C. It is also known as benzine, ligroin, petroleum ether, and naphtha.

* Data from NFPA 497M-1986, *Classification of Gases, Vapors and Dusts for Electrical Equipment in Hazardous (Classified) Locations*.

** Data from NFPA 325M-1984, *Fire Hazard Properties of Flammable Liquids, Gases and Volatile Solids*.

Figure A–1(C) Properties of gases, liquids, and vapors. (Courtesy of Crouse-Hinds, Division of Cooper Industries)

Class II, Group E	Minimum Cloud or Layer Ignition Temp.[1]		
Material[2]	°F		°C
Aluminum, atomized collector fines	1022	Cl	550
Aluminum, A422 flake	608		320
Aluminum — cobalt alloy (60-40)	1058		570
Aluminum — copper alloy (50-50)	1526		830
Aluminum — lithium alloy (15% Li)	752		400
Aluminum — magnesium alloy (Dowmetal)	806	Cl	430
Aluminum — nickel alloy (58-42)	1004		540
Aluminum — silicon alloy (12% Si)	1238	NL	670
Boron, commercial-amorphous (85% B)	752		400
Calcium Silicide	1004		540
Chromium, (97%) electrolytic, milled	752		400
Ferromanganese, medium carbon	554		290
Ferrosilicon (88%, 9% Fe)	1472		800
Ferrotitanium (19% Ti, 74.1% Fe, 0.06% C)	698	Cl	370
Iron, 98%, H_2 reduced	554		290
Iron, 99%, Carbonyl	590		310
Magnesium, Grade B, milled	806		430
Manganese	464		240
Silicon, 96%, milled	1436	Cl	780
Tantalum	572		300
Thorium, 1.2%, O_2	518	Cl	270
Tin, 96%, atomized (2% Pb)	806		430
Titanium, 99%	626	Cl	330
Titanium Hydride, (95% Ti, 3.8% H_2)	896	Cl	480
Vanadium, 86.4%	914		490
Zirconium Hydride, (93.6% Zr, 2.1% H_2)	518		270
Class II, Group G			
AGRICULTURAL DUSTS			
Alfalfa Meal	392		200
Almond Shell	392		200
Apricot Pit	446		230
Cellulose	500		260
Cherry Pit	428		220
Cinnamon	446		230
Citrus Peel	518		270
Cocoa Bean Shell	698		370
Cocoa, natural, 19% fat	464		240
Coconut Shell	428		220
Corn	482		250
Corncob Grit	464		240
Corn Dextrine	698		370
Cornstarch, commercial	626		330
Cornstarch, modified	392		200
Cork	410		210
Cottonseed Meal	392		200
Cube Root, South Amer.	446		230
Flax Shive	446		230
Garlic, dehydrated	680	NL	360
Guar Seed	932	NL	500
Gum, Arabic	500		260
Gum, Karaya	464		240
Gum, Manila (copal)	680	Cl	360
Gum, Tragacanth	500		260
Hemp Hurd	428		220
Lycopodium	590		310
Malt Barley	482		250
Milk, Skimmed	392		200
Pea Flour	500		260
Peach Pit Shell	410		210
Peanut Hull	410		210
Peat, Sphagnum	464		240
Pecan Nut Shell	410		210
Pectin	392		200
Potato Starch, Dextrinated	824	NL	440
Pyrethrum	410		210
Rauwolfia Vomitoria Root	446		230
Rice	428		220
Rice Bran	914	NL	490
Rice Hull	428		220
Safflower Meal	410		210
Soy Flour	374		190
Soy Protein	500		260
Sucrose	662	Cl	350
Sugar, Powdered	698	Cl	370

Class II, Group G (cont'd)	Minimum Cloud or Layer Ignition Temp.		
	°F		°C
AGRICULTURAL DUSTS			
Tung, Kernels, Oil-Free	464		240
Walnut Shell, Black	428		220
Wheat	428		220
Wheat Flour	680		360
Wheat Gluten, gum	968	NL	520
Wheat Starch	716	NL	380
Wheat Straw	428		220
Woodbark, Ground	482		250
Wood Flour	500		260
Yeast, Torula	500		260
CARBONACEOUS DUSTS[3]			
Asphalt, (Blown Petroleum Resin)	950	Cl	510
Charcoal	356		180
Coal, Kentucky Bituminous	356		180
Coal, Pittsburgh Experimental	338		170
Coal, Wyoming	—		—
Gilsonite	932		500
Lignite, California	356		180
Pitch, Coal Tar	1310	NL	710
Pitch, Petroleum	1166	NL	630
Shale, Oil	—		—
CHEMICALS			
Acetoacetanilide	824	M	440
Acetoacet-p-phenetidide	1040	NL	560
Adipic Acid	1022	M	550
Anthranilic Acid	1076	M	580
Aryl-nitrosomethylamide	914	NL	490
Azelaic Acid	1130	M	610
2,2-Azo-bis-butyronitrile	662		350
Benzoic Acid	824	M	440
Benzotriazole	824	M	440
Bisphenol-A	1058	M	570
Chloroacetoacetanilide	1184	M	640
Diallyl Phthalate	896	M	480
Dicumyl Peroxide (suspended on $CaCO_3$), 40-60	356		180
Dicyclopentadiene Dioxide	788	NL	420
Dihydroacetic Acid	806	NL	430
Dimethyl Isophthalate	1076	M	580
Dimethyl Terephthalate	1058	M	570
3,5 - Dinitrobenzoic Acid	860	NL	460
Dinitrotoluamide	932	NL	500
Diphenyl	1166	M	630
Ditertiary Butyl Paracresol	878	NL	470
Ethyl Hydroxyethyl Cellulose	734	NL	390
Fumaric Acid	968	M	520
Hexamethylene Tetramine	770	S	410
Hydroxyethyl Cellulose	770	NL	410
Isotoic Anhydride	1292	NL	700
Methionine	680		360
Nitrosoamine	518	NL	270
Para-oxy-benzaldehyde	716	Cl	380
Paraphenylene Diamine	1148	M	620
Paratertiary Butyl Benzoic Acid	1040	M	560
Pentaerythritol	752	M	400
Phenylbetanaphthylamine	1256	NL	680
Phthalic Anydride	1202	M	650
Phthalimide	1166	M	630
Salicylanilide	1130	M	610
Sorbic Acid	860		460
Stearic Acid, Aluminum Salt	572		300
Stearic Acid, Zinc Salt	950	M	510
Sulfur	428		220
Terephthalic Acid	1256	NL	680
DRUGS			
2-Acetylamino-5-nitrothiazole	842		450
2-Amino-5-nitrothiazole	860		460
Aspirin	1220	M	660
Gulasonic Acid, Diacetone	788	NL	420
Mannitol	860	M	460
Nitropyridone	806	M	430
1-Sorbose	698	M	370
Vitamin B1, mononitrate	680	NL	360
Vitamin C (Ascorbic Acid)	536		280

Figure A-2(A) Properties of combustible dusts. (Courtesy of Crouse-Hinds, Division of Cooper Industries)

Class II, Group G	Minimum Cloud or Layer Ignition Temp.[1]	
	°F	°C
DYES, PIGMENTS, INTERMEDIATES		
Beta-naphthalene-azo-Dimethylaniline	347	175
Green Base Harmon Dye	347	175
Red Dye Intermediate	347	175
Violet 200 Dye	347	175
PESTICIDES		
Benzethonium Chloride	716 Cl	380
Bis(2-Hydroxy-5-chlorophenyl) methane	1058 NL	570
Crag No. 974	590 Cl	310
Dieldrin (20%)	1022 NL	550
2, 6-Ditertiary-butyl-paracresol	788 NL	420
Dithane	356	180
Ferbam	302	150
Manganese Vancide	248	120
Sevin	284	140
α,α - Trithiobis (N,N-Dimethylthio-formamide)	446	230
THERMOPLASTIC RESINS AND MOLDING COMPOUNDS		
Acetal Resins		
Acetal, Linear (Polyformaldehyde)	824	440
Acrylic Resins		
Acrylamide Polymer	464	240
Acrylonitrile Polymer	860	460
Acrylonitrile - Vinyl Pyridine Copolymer	464	240
Acrylonitrile-Vinyl Chloride-Vinylidene Chloride Copolymer (70-20-10)	410	210
Methyl Methacrylate Polymer	824 NL	440
Methyl Methacrylate - Ethyl Acrylate Copolymer	896	480
Methyl Methacrylate-Ethyl Acrylate-Styrene Copolymer	824	440
Methyl Methacrylate-Styrene-Butadiene-Acrylonitrile Copolymer	896 NL	480
Methacrylic Acid Polymer	554	290
Cellulosic Resins		
Cellulose Acetate	644	340
Cellulose Triacetate	806 NL	430
Cellulose Acetate Butyrate	698 NL	370
Cellulose Propionate	860 NL	460
Ethyl Cellulose	608 Cl	320
Methyl Cellulose	644	340
Carboxymethyl Cellulose	554	290
Hydroxyethyl Cellulose	644	340
Chlorinated Polyether Resins		
Chlorinated Polyether Alcohol	860	460
Nylon (Polyamide) Resins		
Nylon Polymer (Polyhexa-methylene Adipamide)	806	430
Polycarbonate Resins		
Polycarbonate	1310 NL	710
Polyethylene Resins		
Polyethylene, High Pressure Process	716	380
Polyethylene, Low Pressure Process	788 NL	420
Polyethylene Wax	752 NL	400
Polymethylene Resins		
Carboxypolymethylene	968 NL	520

Class II, Group G (cont'd)	Minimum Cloud or Layer Ignition Temp.	
	°F	°C
THERMOPLASTIC RESINS AND MOLDING COMPOUNDS		
Polypropylene Resins		
Polypropylene (No Antioxidant)	788 NL	420
Rayon Resins		
Rayon (Viscose) Flock	482	250
Styrene Resins		
Polystyrene Molding Cmpd.	1040 NL	560
Polystyrene Latex	932	500
Styrene-Acrylonitrile (70-30)	932 NL	500
Styrene-Butadiene Latex (> 75% Styrene; Alum Coagulated)	824 NL	440
Vinyl Resins		
Polyvinyl Acetate	1022 NL	550
Polyvinyl Acetate/Alcohol	824	440
Polyvinyl Butyral	734 NL	390
Vinyl Chloride - Acrylonitrile Copolymer	878	470
Polyvinyl Chloride - Dioctyl Phthalate Mixture	608 NL	320
Vinyl Toluene - Acrylonitrile Butadiene Copolymer	936 NL	530
THERMOSETTING RESINS AND MOLDING COMPOUNDS		
Allyl Resins		
Allyl Alcohol Derivative (CR-39)	932 NL	500
Amino Resins		
Urea Formaldehyde Molding Compound	860 NL	460
Urea Formaldehyde - Phenol Formaldehyde Molding Compound (Wood Flour Filler)	464	240
Epoxy Resins		
Epoxy	1004 NL	540
Epoxy - Bisphenol A	950 NL	510
Phenol Furfural	590	310
Phenolic Resins		
Phenol Formaldehyde	1076 NL	580
Phenol Formaldehyde Molding Cmpd (Wood Flour Filler)	932 NL	500
Phenol Formaldehyde, Polyalkylene-Polyamine Modified	554	290
Polyester Resins		
Polyethylene Terephthalate	932 NL	500
Styrene Modified Polyester-Glass Fiber Mixture	680	360
Polyurethane Resins		
Polyurethane Foam, No Fire Retardant	824	440
Polyurethane Foam, Fire Retardant	734	390
SPECIAL RESINS AND MOLDING COMPOUNDS		
Alkyl Ketone Dimer Sizing Compound	320	160
Cashew Oil, Phenolic, Hard	356	180
Chlorinated Phenol	1058 NL	570
Coumarone-Indene, Hard	968 NL	520
Ethylene Oxide Polymer	662 NL	350
Ethylene-Maleic Anhydride Copolymer	1004 NL	540
Lignin, Hydrolized, Wood-Type, Fines	842 NL	450
Petrin Acrylate Monomer	428	220
Petroleum Resin (Blown Asphalt)	932	500
Rosin, DK	734 NL	390
Rubber, Crude, Hard	662 NL	350
Rubber, Synthetic, Hard (33% S)	608 NL	320
Shellac	752 NL	400
Sodium Resinate	428	220
Styrene — Maleic Anhydride Copolymer	878 Cl	470

[1]Normally, the minimum ignition temperature of a layer of a specific dust is lower than the minimum ignition temperature of a cloud of that dust. Since this is not universally true, the lower of the two minimum ignition temperatures is listed. If no symbol appears between the two temperature columns, then the layer ignition temperature is shown. "Cl" means the cloud ignition temperature is shown. "NL" means that no layer ignition temperature is available and the cloud ignition temperature is shown. "M" signifies that the dust layer melts before it ignites; the cloud ignition temperature is shown. "S" signifies that the dust layer sublimes before it ignites; the cloud ignition temperature is shown.

[2]Certain metal dusts may have characteristics that require safeguards beyond those required for atmospheres containing the dusts of aluminum, magnesium, and their commercial alloys. For example, zirconium, thorium, and uranium dusts have extremely low ignition temperatures (as low as 20° C) and minimum ignition energies lower than any material classified in any of the Class I or Class II groups.

[3]The 1987 NEC classifies carbonaceous dusts as Group F, some of which may be conductive.

Figure A–2(B) Properties of combustible dusts. (Courtesy of Crouse-Hinds, Division of Cooper Industries)

USES FOR RACEWAY AND OTHER WIRING METHODS
CHAPTER 3, NEC®

NEC ARTICLE	TYPE OF WIREWAY	For conductors over 600 volts	Inside Buildings	Outside Buildings	Under Ground	Cinder Fill	Embedded in Concrete	Wet locations	Dry Locations	Corrosive Locations	Severe Corrosive Locations	Hazardous Locations	Mechanical Injury	Severe Mechanical injury	Exposed work	Concealed work
330	TYPE MI MINERAL INSULATED-SHEATHED CABLE	X	P	P	C	X	P	P	P	C	X	P	P	P	P	P
331	ELECTRICAL NONMETALLIC TUBING*	X	C	X	X	X	C	P	P	P	C	X	X	X	C	C
333	TYPE AC ARMORED CABLE	X	P	X	X	X	X	X	P	X	X	X	X	X	C	P
334	TYPE MC METAL CLAD CABLE	C	C	C	C	X	C	C	P	X	X	C	X	X	P	P
336	NONMETALLIC SHEATHED CABLE ROMEX TYPE NM	X	P	C	X	X	X	X	P	X	X	X	X	X	P	P
338	TYPE SE SERVICE-ENTRANCE CABLE	X	C	P	X	X	X	C	C	X	X	X	X	X	C	C
388	TYPE USE SERVICE-ENTRANCE CABLE	X	X	C	P		X	C	X	X	X	X	X	X	X	X,
339	TYPE UF UNDERGROUND FEEDER AND BRANCH CIRCUIT CABLE	X	P	P	P		X	P	P	X	X	X	X	X	C	C
343	PREASSEMBLED CABLE IN NONMETALLIC CONDUIT	P	X	-	P	P		C	X	C	C	X	-	-	-	-
345	IMC INTERMEDIATE METAL CONDUIT	P	P	P	P	C	P	P	P	P	C	P	P	P	P	P
345	IMC INTERMEDIATE COATED METAL CONDUIT	P	C	P	P	P	P	P	P	P	P	P	P	P	P	P
346	RIGID METAL CONDUIT (STEEL)	P	P	P	P	C	P	P	P	P	C	P	P	P	P	P
346	RIGID COATED METAL CONDUIT	P	C	P	P	P	P	P	P	P	P	P	P	P	P	P
346	RIGID METAL CONDUIT (ALUMINUM)	P	P	P	X	X	X	P	P	X	X	P	P	C	P	P
347	RIGID NONMETALLIC CONDUIT O (SCHEDULE 40) *	C	P	P	P	P	P	P	P	P	P	X	X	X	C	P
347	RIGID NONMETALLIC CONDUIT (SCHEDULE 80) *	C	P	P	P	P	P	P	P	P	P	X	P	P	P	P
348	EMT ELECTRICAL METALLIC TUBING	C	P	P	P	C	P	P	P	C	P	C	P	X	P	P
349	FLEXIBLE METALLIC TUBING	C	C	X	X	X	X	X	C	X	X	X	X	X	C	C
350	FLEXIBLE METAL CONDUIT STEEL AND ALUMINUM	X	P	C	X	X	X	C	P	X	X	X	X	X	P	P
351A	LIQUIDTIGHT FLEXIBLE METAL CONDUIT	X	P	P	C	C	C	C	P	X	X	C	X	X	P	P
351b	LIQUIDTIGHT FLEXIBLE NONMETALLIC CONDUIT *	X	P	C	C	C	C	C	P	P	P	C	X	X	P	P
352A	SURFACE METAL RACEWAYS	C	P	C	-	-	-	C	P	C	C	C	C	C	P	-
352B	SURFACE NONMETALLIC RACEWAYS *	C	P	C	-	-	-	C	P	C	C	C	C	X	P	-
354	UNDERFLOOR RACEWAYS	C	P	-	-	-	C	X	P	X	X	C	C	C	-	P
356	CELLULAR METAL FLOOR RACEWAYS	C	P	-	-	-	P	X	P	-	-	—	-	-	-	-
362A	WIREWAYS METAL	C	P	C	X	X	X	P	P	X	X	C	P	X	P	X
362B	WIREWAYS NONMETALLIC	C	C	X	X	X	X	C	C	C	C	X	C	X	P	X
364	BUSWAYS	C	P	C	X	X	X	C	P	X	X	X	P	X	P	C

P - GENERALLY PERMITTED   C - CONDITIONAL   X - NOT PERMITTED   — CONDITIONS DO NOT APPLY

Wiring methods covered by *Chapter 3, NEC®* are limited to systems utilizing voltages not exceeding 600 volts unless higher voltages are specifically permitted. *NEC® 300–21 & 22* may further limit some wiring methods.

* - Where subject to chemicals for which the raceway is specifically approved
* - (FPN) Extreme cold may cause some nonmetallic conduits to become brittle and therefore more susceptible to damage from physical contact
* - Except where the enclosed conductors' ambient temperature would exceed those for which the conduit is approved

Figure A–3 Chart showing recommended uses or applications for all types of wiring methods.

WEIGHTS AND DIMENSIONS FOR GALVANIZED RIGID CONDUIT

Trade Size, Inches	Approx. Wt. per 100 ft. (30.5m)		Nominal Outside Dia.[1]		Nominal Wall Thickness[2]		Length of Finished Conduit[3]		Quantity In Primary Bundle		Quantity In Master Bundle		Approx. Wt. of Master Bundle		Volume of Master Bundle	
	lb.	kg	in.	mm	in.	mm	ft.	m	ft.	m	ft.	m	lb.	kg	cu ft.	cu m
½	80	36.29	0.840	21.3	0.104	2.6	9'11¼"	3.03	100	30.48	2500	762	2000	907	20.8	0.59
¾	109	49.44	1.050	26.7	0.107	2.7	9'11¼"	3.03	50	15.24	2000	610	2180	989	24.3	0.69
1	165	74.84	1.315	33.4	0.126	3.2	9'11"	3.02	50	15.24	1250	381	2063	936	21.7	0.61
1¼	215	97.52	1.660	42.2	0.133	3.4	9'11"	3.02	30	9.14	900	274	1935	878	23.3	0.66
1½	258	117.03	1.900	48.3	0.138	3.5	9'11"	3.02	—	—	800	244	2064	936	27.8	0.79
2	352	159.67	2.375	60.3	0.146	3.7	9'11"	3.02	—	—	600	183	2112	958	33.8	0.96
2½	567	257.19	2.875	73.0	0.193	4.9	9'10½"	3.01	—	—	370	113	2098	952	29.2	0.83
3	714	323.87	3.500	88.9	0.205	5.2	9'10½"	3.01	—	—	300	91	2142	972	31.3	0.89
3½	860	390.10	4.000	101.6	0.215	5.5	9'10¼"	3.00	—	—	250	76	2150	975	34.7	0.98
4	1000	453.60	4.500	114.3	0.225	5.7	9'10¼"	3.00	—	—	200	61	2000	907	33.7	0.95
5	1320	598.75	5.563	141.3	0.245	6.2	9'10"	3.00	—	—	150	46	1980	898	41.3	1.17
6	1785	809.68	6.625	168.3	0.266	6.8	9'10"	3.00	—	—	100	30	1785	810	38.9	1.10

[1]Outside diameter tolerances: +/- .015 in. (.38mm) for trade sizes ½ in. through 2 in. +/- .025 in. (.64mm) for trade sizes 2½ in. through 4 in. +/- 1% for trade sizes 5 in. and 6 in.
[2]For more information only; not a spec requirement.
[3]Without Coupling Length Tolerances: +/- .25 in (6.35mm).

Figure A–4 Galvanized rigid conduit specifications. (Courtesy of Allied Tube & Conduit)

WEIGHTS AND DIMENSIONS FOR INTERMEDIATE METAL CONDUIT

Trade Size, Inches	Approx. Wt. per 100 ft. (30.5m)		Nominal Outside Dia.[1]		Minimum Wall Thickness[2]		Length of Finished Conduit[3]		Quantity In Primary Bundle		Quantity In Master Bundle		Approx. Wt. of Master Bundle		Volume of Master Bundle	
	lb.	kg	in.	mm	in.	mm	ft.	m	ft.	m	ft.	m	lb.	kg	cu ft.	cu m
½	60	27.22	0.815	20.7	0.070	1.8	9'11¼"	3.03	100	30.48	3500	1067	2233	1013	26.7	0.76
¾	82	37.29	1.029	26.1	0.075	1.9	9'11¼"	3.03	50	15.24	2500	762	2078	943	30.7	0.87
1	116	52.62	1.290	32.8	0.085	2.2	9'11"	3.02	50	15.24	1700	518	2035	923	30.7	0.87
1¼	150	68.04	1.638	41.6	0.085	2.2	9'11"	3.02	—	—	1350	411	2209	1002	36.3	1.03
1½	182	82.55	1.883	47.8	0.090	2.3	9'11"	3.02	—	—	1100	335	2122	963	38.2	1.08
2	242	109.77	2.360	59.9	0.095	2.4	9'11"	3.02	—	—	800	244	2096	951	45.8	1.30
2½	428	194.14	2.857	72.6	0.140	3.5	9'10½"	3.01	—	—	370	113	1652	749	29.2	0.83
3	526	238.59	3.476	88.3	0.140	3.5	9'10½"	3.01	—	—	300	91	1618	734	31.3	0.89
3½	612	277.60	3.971	100.9	0.140	3.5	9'10¼"	3.00	—	—	240	73	1576	715	34.7	0.98
4	682	309.35	4.466	113.4	0.140	3.5	9'10¼"	3.00	—	—	240	73	1809	821	42.8	1.21

[1]Outside diameter tolerances: +/- .005 in. (.13mm) for trade sizes ½" through 1". +/- .0075 in. (.19mm) for trade size 1¼". through 2 in. +/- .010 in. (.25mm) for trade size 2½" through 4 in.
[2]Wall thickness tolerances: + .015 in. (.38mm) and - .000 for trade sizes ½" through 2 in. + 0.20 in. (.51 mm) and -.000 for trade sizes 2½" through 4 in.
Without Coupling Length Tolerances: +/- .25 in (6.35mm).

Figure A–5 Intermediate metal conduit specifications. (Courtesy of Allied Tube & Conduit)

WEIGHTS AND DIMENSIONS FOR ELECTRICAL METALLIC TUBING

Trade Size, Inches	Approx. Wt. per 100 ft. (30.5m)		Nominal Outside Dia.[1]		Nominal Wall Thickness		Length of Finished Conduit[2]		Quantity In Primary Bundle		Quantity In Master Bundle		Approx. Wt. of Master Bundle		Volume of Master Bundle	
	lb.	kg	in.	mm	in.	mm	ft.	m	ft.	m	ft.	m	lb.	kg	cu ft.	cu m
½	29	13.15	0.706	17.9	0.042	1.067	10	3.05	100	30.48	7000	2134	2037	924	28.7	0.81
¾	45	20.41	0.922	23.4	0.049	1.245	10	3.05	100	30.48	5000	1524	2175	987	35.6	1.01
1	65	29.48	1.163	29.5	0.057	1.448	10	3.05	50	15.24	3000	914	1905	864	33.7	0.95
1¼	96	43.55	1.510	38.4	0.065	1.651	10	3.05	50	15.24	2000	610	1894	859	35.0	0.99
1½	111	50.35	1.740	44.2	0.065	1.651	10	3.05	50	15.24	1500	457	1692	767	34.2	0.97
2	141	63.96	2.197	55.8	0.065	1.651	10	3.05	—	—	1200	366	1693	768	46.7	1.32
2½	215	97.52	2.875	73.0	0.072	1.829	10	3.05	—	—	610	186	1412	640	41.5	1.18
3	260	117.94	3.500	88.9	0.072	1.829	10	3.05	—	—	510	155	1429	648	48.9	1.38
3½	325	147.42	4.000	101.6	0.083	2.108	10	3.05	—	—	370	113	1248	566	48.6	1.38
4	390	176.90	4.500	114.3	0.083	2.108	10	3.05	—	—	300	91	1134	514	48.3	1.37

[1] Outside diameter tolerances: +/- .005 in. (.13mm) for trade sizes ½" through 2". +/- .010 in. (.25mm) for trade size 2½". +/- .015 in. (.38mm) for trade size 3". +/- .020 in. (.51mm) for trade sizes 3½" and 4".
[2] Length tolerances: +/-.25" (6.35mm).

Figure A-6 Electrical metallic tubing (EMT) specifications. (Courtesy of Allied Tube & Conduit)

PLUS 40® Heavy Wall

Nom. Size	Part No.	O.D.	I.D.	Wall	Wt. Per 100 Feet	Feet Per Bundle
½	49005	.840	.622	.109	18	100
¾	49007	1.050	.824	.113	23	100
1	49008	1.315	1.049	.133	35	100
1¼	49009	1.660	1.380	.140	48	50
1½	49010	1.900	1.610	.145	57	50
2	49011	2.375	2.067	.154	76	50
2½	49012	2.875	2.469	.203	125	10
3	49013	3.500	3.066	.216	164	10
3½	49014	4.000	3.548	.226	198	10
4	49015	4.500	4.026	.237	234	10
5	49016	5.563	5.047	.258	317	10
6	49017	6.625	6.065	.280	412	10

Rigid nonmetallic conduit is normally supplied in standard 10' lengths, with one belled end per length. For specific requirements, it may be produced in lengths shorter or longer than 10', with or without belled ends.

Figure A-7 Rigid nonmetallic conduit (Schedule 40) specifications. (Courtesy of Carlon, a Lampson Sessions Company)

PLUS 80® Extra Heavy Wall

Nom. Size	Part No.	O.D.	I.D.	Wall	Wt. Per 100 Feet	Feet Per Bundle
½	49405	.840	.546	.147	21	100
¾	49407	1.050	.742	.154	29	100
1	49408	1.315	.957	.179	43	100
1¼	49409	1.660	1.278	.191	60	50
1½	49410	1.900	1.500	.200	72	50
2	49411	2.375	1.939	.218	100	10
2½	49412	2.875	2.323	.276	153	10
3	49413	3.500	2.900	.300	212	10
4	49415	4.500	3.826	.337	310	10
5	49416	5.563	4.813	.375	431	10
6	49417	6.625	6.193	.432	592	10

Rigid nonmetallic conduit is normally supplied in standard 10' lengths, with one belled end per length. For specific requirements, it may be produced in lengths shorter or longer than 10', with or without belled ends.

Figure A-8 Rigid nonmetallic conduit (Schedule 80) specifications. (Courtesy of Carlon, a Lampson Sessions Company)

Approximate Diameters for Conduit, Locknuts, and Bushings

Component	Nominal or trade size of conduit, inches											
Approximate diameter, inches	1/2	3/4	1	1 1/4	1 1/2	2	2 1/2	3	3 1/2	4	5	6
Conduit	7/8	1 1/16	1 3/8	1 11/16	1 15/16	2 3/8	2 7/8	3 1/2	4	4 1/2	5 9/16	6 5/8
Locknut	1 1/8	1 3/8	1 11/16	2 3/16	2 7/16	3	3 7/16	4 3/16	5 3/8	6 11/16	7 15/16	
Bushing	1	1 1/4	1 1/2	1 7/8	2 1/8	2 5/8	3 3/16	3 7/8	4 7/16	5	6 1/4	7 3/8

Figure A–9 Approximate dimensions of standard locknuts and bushings. (Courtesy of American Iron and Steel Institute)

Recommended Minimum Spacing of Rigid Steel Conduit and EMT at Junction and Pull Boxes

Nominal or trade size inches	Distances between centers, inches											
	1/2	3/4	1	1 1/4	1 1/2	2	2 1/2	3	3 1/2	4	5	6
1/2	1 3/8	—	—	—	—	—	—	—	—	—	—	—
3/4	1 1/2	1 5/8	—	—	—	—	—	—	—	—	—	—
1	1 3/4	1 7/8	2	—	—	—	—	—	—	—	—	—
1 1/4	2	2 1/8	2 1/4	2 1/2	—	—	—	—	—	—	—	—
1 1/2	2 1/8	2 1/4	2 3/8	2 5/8	2 3/4	—	—	—	—	—	—	—
2	2 3/8	2 1/2	2 3/4	3	3 1/8	3 3/8	—	—	—	—	—	—
2 1/2	2 5/8	2 3/4	3	3 1/4	3 3/8	3 5/8	4	—	—	—	—	—
3	3	3 1/8	3 3/8	3 5/8	3 3/4	4	4 3/8	4 3/4	—	—	—	—
3 1/2	3 3/8	3 1/2	3 5/8	3 7/8	4	4 3/8	4 5/8	5	4 3/4	—	—	—
4	3 11/16	3 7/8	4	4 1/4	4 3/8	4 3/4	5	5 3/8	5 5/8	6	—	—
5	4 3/8	4 1/2	4 5/8	4 7/8	5	5 3/8	5 5/8	6	6 1/4	6 5/8	7 1/4	—
6	5	5 1/8	5 1/4	5 1/2	5 5/8	6	6 1/4	6 5/8	7	7 1/4	8	8 5/8

Figure A–10 Recommended spacing for conduit and tubing where run in racks or where entering a panelboard or junction box. (Courtesy of American Iron and Steel Institute)

Basic Electrical Formulas

DC Circuit Characteristics

Ohm's Law:

$$E = IRI = \frac{E}{R}R = \frac{E}{I}$$

E = voltage impressed on circuit (volts)
I = current flowing in circuit (amperes)
R = circuit resistance (ohms)

Resistances in Series:

$R_t = R_1 + R_2 + R_3 + \ldots$
R_T = total resistance (ohms)
R_1, R_2 etc. = individual resistances (ohms)

Resistances in Parallel:

$$R_t = \frac{1}{\frac{1}{R_1} + \frac{1}{R_2} + \frac{1}{R_3} + \ldots}$$

Formulas for the conversion of electrical and mechanical power:

$HP = \frac{watts}{746}$ (watts × .00134)

$HP = \frac{kilowatts}{.746}$ (kilowatts × 1.34)

Kilowatts = HP × .746
Watts = HP × 746
HP = Horsepower

In direct-current circuits, electrical power is equal to the product of the voltage and current:

$$P = EI = I_2R = \frac{E^2}{R}$$

P = power (watts)
E = voltage (volts)
I = current (amperes)
R = resistance (ohms)

Solving the basic formula for I, E, and R gives

$$I = \frac{P}{E} = \sqrt{\frac{P}{R}}; \quad E = \frac{P}{I} = \sqrt{RP}; \quad R = \frac{E^2}{P} = \frac{P}{I^2}$$

Energy

Energy is the capacity for doing work. Electrical energy is expressed in kilowatt-hours (kWhr), one kilowatt-hour representing the energy expended by a power source of 1 kW over a period of 1 hour.

Efficiency

Efficiency of a machine, motor or other device is the ratio of the energy output (useful energy delivered by the machine) to the energy input (energy delivered to the machine), usually expressed as a percentage:

$$\text{Efficiency} = \frac{\text{output}}{\text{input}} \times 100\%$$

or Output = imput × $\frac{\text{efficiency}}{100\%}$

Torque

Torque may be described as a force tending to cause a body to rotate. It is expressed in pound-feet or pounds of force acting at a certain radius:

Torque (pound-feet) = force tending to produce rotation (pounds) × distance from center of rotation to point at which force is applied (feet).

Relations between torque and horsepower:

$$\text{Torque} = \frac{33{,}000 \times HP}{6.28 \times rpm}$$

$$HP = \frac{6.28 \times rpm \ time \ torque}{33{,}000}$$

rpm = speed of rotating part (revolutions per minute)

AC Circuit Characteristics

The instantaneous values of an alternating current or voltage vary from zero to maximum value each half cycle. In the practical formula that follows, the "effective value" of current and voltage is used, defined as follows:

Effective value = 0.707 × maximum instantaneous value

Inductances in Series and Parallel:

The resulting circuit inductance of several inductances in series or parallel is determined exactly as the sum of resistances in series or parallel as described under DC circuit characteristics.

Impedance:

Impedance is the total opposition to the flow of alternating current. It is a function of resistance, capacitive reactance and inductive reactance. The following formulae relate these circuit properties:

$$X_L = 2\pi HzL \quad X_c = \frac{1}{2\pi HzC} \quad Z = \sqrt{R^2 + (X_L - X_C)^2}$$

X_L = inductive reactance (ohms)
X_C = capacitive reactance (ohms)
Z = impedance (ohms)
Hz - (Hertz) cycles per second
C = capacitance (farads)
L - inductance (henrys)
R = resistance (ohms)
π = 3.14

In circuits where one or more of the properties L, C, or R is absent, the impedance formula is simplified as follows:

Resistance only:	Inductance only:	Capacitance only:
Z = R	Z = X_L	Z = XC
Resistance and Inductance only:	Resistance and Capacitance only:	Inductance and Capacitance only:
Z = $\sqrt{R^2 + X_L^2}$	Z = $\sqrt{R^2 + X_C^2}$	Z = $\sqrt{X_L - X_C^2}$

Ohm's law for AC circuits:

$E = 1 \times Z \qquad I = \frac{E}{Z} \qquad Z = \frac{E}{I}$

Capacitances in Parallel:

$C_t = C_1 + C_2 \ C_3 + \ldots$
C_t = total capacitance (farads)
$C_1 C_2 C_3 \ldots$ = individual cpacitances (farads)

Figure A–11(A) Common electrical formulas for making AC and DC calculations. (Courtesy of American Iron and Steel Institute)

Capacitances in series:

$$C_t = \dfrac{1}{\dfrac{1}{C_1} + \dfrac{1}{C_2} + \dfrac{1}{C_3} + \ldots}$$

Phase Angle

An alternating current through an inductance lags the voltage across the inductance by an angle computed as follows:

Tangent of angle of lag $= \dfrac{X_L}{R}$

An alternating current through a capacitance leads the voltage across the capacitance by an angle computed as follows:

Tangent of angle of lead $= \dfrac{X_C}{R}$

The resultant angle by which a current leads or lags the voltage in an entire circuit is called the phase angle and is computed as follows:

Cosine of phase angle $= \dfrac{R \text{ of circuit}}{Z \text{ of circuit}}$

Power Factor

Power factor of a circuit or system is the ratio of actual power (watts) to apparent power (volt-amperes), and is equal to the cosine of the phase angle of the circuit:

$PF = \dfrac{\text{actual power}}{\text{apparent power}} = \dfrac{\text{watts}}{\text{volts} \times \text{amperes}} = \dfrac{kW}{kVA} = \dfrac{R}{Z}$

KW = kilowatts
kVA = kilowatt-amperes = volt-amperes ÷ 1,000
PF = power factor (expressed as decimal or percent)

Single-Phase Circuits

$kVA = \dfrac{EI}{1,000} = \dfrac{kW}{PF}$ kW = kVA × PF

$I = \dfrac{P}{E \times PF}$ $E = \dfrac{P}{I \times PF}$ $PF = \dfrac{P}{E \times I}$

P = E × I × PF
P = power (watts)

Two-Phase Circuits

$I = \dfrac{P}{2 \times E \times PF}$ $E = \dfrac{P}{2 \times I \times PF}$ $PF = \dfrac{P}{E \times I}$

$kVA = \dfrac{2 \times E \times I}{1000} = \dfrac{kW}{PF}$ kW = kVA × PF

P = 2 × E × I × PF
E = phase voltage (volts)

Three-Phase Circuits, Balanced Star or Wye

$I_N = 0$ $I = I_P E = \sqrt{3} E_P = 1.73 E_P$

$E_P = \dfrac{E}{\sqrt{3}} = \dfrac{E}{1.73} = 0.577 E$

I_N = current in neutral (amperes)
I = line current per phase (amperes)
I_P = current in each phase winding (amperes)
E = voltage, phase to phase (volts)
E_P = voltage, phase to neutral (volts)

Three-Phase Circuits, Balanced Delta

$I = 1.732 \times I_P I_P = \dfrac{1}{\sqrt{3}} = 0.577 \times I$

$E = E_P$

Power:

Balanced 3-Wire, 3-Phase Circuit, Delta or Wye

For unit power factor (PF = 1.0):
P = 1.732 × E × I

$I = \dfrac{P}{\sqrt{3} E} = 0.577 \dfrac{P}{E}$ $E = \dfrac{P}{\sqrt{3} I} = 0.577 \dfrac{P}{I}$

P = total power (watts)
For any load:
P = 1.732 × E × I × PF VA = 1.732 × E × I

$E = \dfrac{P}{PF \times 1.73 \times I} = 0.577 \times \dfrac{P}{PF \times I}$

$I = \dfrac{P}{PF \times 1.73 \times E} = 0.577 \times \dfrac{P}{I} \times E$

$PF = \dfrac{P}{1.73 \times I \times E} = \dfrac{0.577 \times P}{I \times E}$

VA = apparent power (volt-amperes)
P = actual power (watts)
E = line voltage (volts)
I = line current (amperes)

Power Loss:

Any AC or DC Circuit

$P = I^2 R I = \sqrt{\dfrac{P}{R}} R = \dfrac{P}{I^2}$

P = power heat loss in circuit (watts)
I = effective current in conductor (amperes)
R = conductor resistance (ohms)

Load Calculations

Branch Circuits—Lighting & Appliance 2-Wire:

$I = \dfrac{\text{total connected load (watts)}}{\text{line voltage (volts)}}$

I = current load on conductor (amperes)

3-Wire:

Apply same formula as for 2-wire branch circuit, considering each line to neutral separately. Use line-to-neutral voltage; result gives current in line conductors.

Figure A–11(B) Common electrical formulas for making AC and DC calculations. (Courtesy of American Iron and Steel Institute)

Electrical Wiring Symbols
Selected from American National Standard Graphic for
Electrical Wiring and Layout Diagrams Used in Architecture and Building Construction
ANSI Y32.9-1972

1. Lighting Outlets

	Ceiling	Wall
1.1 Surface or pendant incandescent, mercury-vapor, or similar lamp fixture	○	—○
1.2 Recessed incandescent, mercury-vapor or similar lamp fixture	(R)	—(R)
1.3 Surface of pendant individual fluorescent fixture	[○]	—[○]
1.4 Recessed individual fluorescent fixture	[○R]	—[○R]
1.5 Surface or pendant continuous-row fluorescent fixture	[○——]	
1.6 Recessed continuous-row fluorescent fixture	[○R——]	
1.8 Surface or pendant exit light	(X)	—(X)
1.9 Recessed exit light	(RX)	—(RX)
1.10 Blanked outlet	(B)	—(B)
1.11 Junction box	(J)	—(J)
1.12 Outlet controlled by low-voltage switching when relay is installed in outlet box	(L)	—(L)

2. Receptacle Outlets

	Grounded	Ungrounded
2.1 Single receptacle outlet	—⊖	—⊖$_{UNG}$
2.2 Duplex receptacle outlet	=⊖	=⊖$_{UNG}$
2.3 Triplex receptacle outlet	=⊕	=⊕$_{UNG}$
2.4 Quadrex receptacle outlet	=⊕	=⊕$_{UNG}$
2.5 Duplex receptacle outlet – split wired	=⊖	=⊖$_{UNG}$
2.6 Triplex receptacle outlet – split wired	=⊕	=⊕$_{UNG}$
2.7 Single special-purpose receptacle outlet – split wired	—△	—△$_{UNG}$
2.8 Duplex special-purpose receptacle outlet See note 2.7A	=△	=△$_{UNG}$

NOTE 2.7A: Use numeral or letter as a subscript alongside the symbol, keyed to explanation in the drawing list of symbols, to indicate type of receptacle or usage.

Figure A–12(A) Common electrical symbols. (Courtesy of American Iron and Steel Institute)

2.9 Range outlet (typical)
See note 2.7A

2.10 Special-purpose connection or provision for connection
Use subscript letters to indicate function (SW – dishwasher; CD – clothes dryer, etc).

2.12 Clock hanger receptacle

2.13 Fan hanger receptacle

2.14 Floor single receptacle outlet

2.15 Floor duplex receptacle outlet

2.16 Floor special-purpose outlet
See note 2.7A

2.17 Floor telephone outlet – public

2.18 Floor telephone outlet – private

2.19 Underfloor duct and junction box for triple, double, or single duct system (as indicated by the number of parallel lines)

2.20 Cellular floor header duct

3. Switch Outlets

3.1 Single-pole switch

S

3.2 Double-pole switch

S_2

3.3 Three-way switch

S_3

3.4 Four-way switch

S_4

3.5 Key-operated switch

S_K

3.6 Switch and pilot lamp

S_P

3.7 Switch for low-voltage switching system

S_L

3.8 Master switch for low-voltage switching system

S_{LM}

3.9 Switch and single receptacle

3.10 Switch and double receptacle

3.11 Door switch

S_D

3.12 Time switch

S_T

3.13 Circuit breaker switch

S_{CB}

Figure A–12(B) Common electrical symbols. (Courtesy of American Iron and Steel Institute)

3.14 Momentary contact switch or pushbutton for other than signalling system

SMC

3.15 Ceiling pull switch

(S)

5.13 Radio outlet

[R]

5.14 Television outlet

[TV]

6. Panelboards, switchboards, and related equipment

6.1 Flush-mounted panel board and cabinet
NOTE 6.1A: Identify by notation or schedule

6.2 Surface-mounted panel board and cabinet
See note 6.1A

6.3 Switchboard, power control center, unit substations (should be drawn to scale)
See note 6.1A

6.4 Flush-mounted terminal cabinet
See note 6.1A

NOTE 6.4A: In small-scale drawings the TC may be indicated alongside the symbol

TC

6.5 Surface-mounted terminal cabinet
See note 6.1A and 6.4A

TC

6.6 Pull box
Identify in relation to wiring system section and size

6.7 Motor or other power controller
See note 6.1A

MC

6.8 Externally operated disconnection switch
See note 6.1A

6.9 Combination controller and disconnection means
See note 6.1A

7. Bus Ducts and Wireways

7.1 Trolley duct
See note 6.1A

| T | | T | | T |

7.2 Busway (service, feeder, or plug-in)
See note 6.1A

| B | | B | | B |

7.3 Cable through, ladder, or channel
See note 6.1A

| BP | | BP | | BP |

7.4 Wireway
See note 6.1A

| W | | W | | W |

9. Circuiting

Wiring method indentification by notation on drawing or in specifications.

9.1 Wiring concealed in ceiling or wall

NOTE 9.1A: Use heavy weight line to identify service and feed runs.

9.2 Wiring concealed in floor
See note 9.1A

Figure A–12(C) Common electrical symbols. (Courtesy of American Iron and Steel Institute)

9.2 Wiring exposed
   See note 9.1A

   — — — — — — — — —

9.4 Branch circuit home run to panel board
   Number of arrows indicates number of circuits. (A numeral at each arrow may be used to identify circuit number.)

   ⎯⎯⎯⎯⎯⎯⎯⎯⎯⎯→² →¹

NOTE: Any circuit without further identification indicates a 2-wire circuit. For a greater number of wires, indicate with cross lines (see 9.4.1, Applications)

9.4.1 Applications:

   ⎯⎯⎯///⎯⎯⎯ 3 wires;
   ⎯⎯⎯////⎯⎯⎯ 4 wires, etc

Unless indicated otherwise, the wire size of the circuit is the minimum size required by the specification. Indicate size in inches and identify different functions of wiring system, such as signaling, by notation or other means.

9.6 Wiring turned up

   ⎯⎯⎯⎯⎯⎯⎯⎯○

9.7 Wiring turned down

   ⎯⎯⎯⎯⎯⎯⎯⎯●

Figure A–12(D)   Common electrical symbols. (Courtesy of American Iron and Steel Institute)

These environmental resistance ratings are based upon tests where the specimens were placed in complete submergence in the reagent listed. In many applications Carflex* conduit can be used in process areas where these chemicals are manufactured or used because worker safety requirements dictate that any air presence or splashing be at a very low level. Most liquidtight conduit is located in areas suitable for worker access.

If there are any questions for specific suitability in a given environment, prototype samples should be tested under actual conditions.

RATING CODE

A-Excellent service.
No harmful effect to reduce service life. Suitable for continuous service.

B-Good service life.
Moderate to minor effect. Good for intermittent service. Generally suitable for continuous service.

C-Fair or limited service.
Depends on operating conditions. Generally suitable for intermittent continuous service.

D-Unsatisfactory service.
Not recommended.

*These ratings apply only to Carflex® conduit & tubing.

Chemical	Concentration	Resistance	Chemical	Concentration	Resistance	Chemical	Concentration	Resistance
Acetate Solvents		D	Creosote		D	Methyl Acetate		D
Acetic Acid	10%	B	Cresol		C	Methyl Alcohol		C
Acetic Acid (Glacial)		C	Cresylic Acid		D	Methyl Bromide		D
Acetone		D	Cyclohexane		B	Methyl Ethyl Ketone		D
Acrylonitrile		A	DDT Weed Killer		A	Methylene Chloride		D
Alcohols (Aliphatic)		C	Dibutyl Phthalate		D	Mineral Oil		A
Aluminum Chloride		A	Diesel Oils		C	Monochlorobenzene		D
Aluminum Sulfate (Alums)		A	Diethylene Glycol		B	Muriatic Acid (See Hydrochloric Acid)		—
Ammonia (Anhydrous Liquids)		D	Diethyl Ether		A	Naphtha		C
Ammonia (Aqueous)		A	Di-isodecyl Phthalate		D	Naphthalene		D
Ammoniated Latex		A	Dioctyl Phthalate		D	Nitric Acid	10%	A
Ammonium Chloride		A	Dow General Weed Killer (Phenol)		D	Nitric Acid	35%	A
Ammonium Hydroxide		A	Dow General Weed Killer (H$_2$O)		B	Nitric Acid	70%	D
Amyl Acetate		D	Ethyl Alcohol		C	Oleic Acid		A
Aniline Oils		D	Ethylene Dichloride		D	Oleum		D
Aromatic Hydrocarbons		D	Ethylene Glycol		B	Oxalic Acid		A
Asphalt		D	Ferric Chloride		A	Pentachlorophenol in Oil		B
ASTM Fuel A		C	Ferric Sulfate		A	Pentane		C
ASTM Fuel B		D	Ferrous Chloride		A	Perchloroethylene		D
ASTM #1 Oil		B	Ferrous Sulfate		A	Petroleum Ether		C
ASTM #3 Oil		C	Formaldehyde		D	Phenol		B
Barium Chloride		A	Fuel Oil		B	Phosphoric Acid	85%	A
Barium Sulfide		A	Furfural		C	Pitch		B
Barium Hydroxide		A	Gallic Acid		A	Potassium Hydroxide		A
Benzene (Benzol)		D	Gasoline (Hi Test)		C	Propyl Alcohol		B
Benzine (Petroleum Ether)		C	Glycerine		A	Ritchfield "A" Weed Killer		C
Black Liquor		A	Grease		A	Sea Water		A
Bordeaux Mixture		A	Green Sulfate Liquor		A	Sodium Hydroxide	10%	A
Boric Acid		A	Heptachlor in Petroleum Solvents		A	Sodium Hydroxide	50%	A
Butyl Acetate		D	Heptane		C	Soybean Oil		C
Butyl Alcohol		B	Hexane		C	Sodium Cyanide		A
Calcium Hydroxide		A	Hydrobromic Acid		A	Stoddard Solvent		D
Calcium Hypochlorite		A	Hydrochloric Acid	10%	A	Styrene		D
Carbolic Acid (Phenol)		B	Hydrochloric Acid	40%	C	Sulfur Dioxide (liquid)		D
Carbon Dioxide		A	Hydrofluoric Acid	70%	D	Sulfuric Acid	50%	A
Carbon Disulfide		D	Hydrofluoroboric Acid		A	Sulfuric Acid	98%	D
Carbon Tetrachloride		D	Hydrofluorosilicic Acid		A	Sulfurous Acid		B
Carbonic Acid		A	Hydrogen Peroxide	10%	A	Tall Oil		D
Casein		A	Iso-octane		C	Tannic Acid		A
Caustic Soda		A	Isopropyl Acetate		D	Toluene		D
Chlorine Gas (wet)		D	Isopropyl Alcohol		B	Trichlorethylene		D
Chlorine Gas (dry)		D	Jet Fuels (JP-3, 4, and 5)		C	Triethanol Amine		C
Chlorine (water solution)		C	Kerosene		C	Tricresyl Phosphate (Skydrol)		D
Chlorobenzene		D	Ketones		D	Turpentine		C
Chlorinated Hydrocarbons		D	Linseed Oil		A	Vinegar		A
Chromic Acid	10%	B	Lubricating Oils		A	Vinyl Chloride		D
Citric Acid		A	Magnesium Chloride		A	Water		A
Coal Tar		D	Magnesium Hydroxide		A	White Liquor		A
Coconut Oil		C	Magnesium Sulfate		A	Xylene		D
Corn Oil		A	Malathion 50 in Aromatics		D	Zinc Chloride		A
Cottonseed Oil		C	Malic Acid		A	Zinc Sulfate		A

Figure A–13 Chemical resistance chart and recommended rating code for PVC electrical products with specific liquids and dusts. (Courtesy of Carlon, a Lampson Sessions Company)

Urethane Interior Coating Chemical Resistance Chart

Solutions	Conc.	Temp.	Recommended Exposure		
			Spashing	Liquid	Fumes
Acetic Acid	10%	75	yes	yes	yes
Acid Copper Plating Solution		75	yes	yes	yes
Alkaline Cleaners		75	yes	yes	yes
Aluminum Chloride	Sat'd	75	yes	yes	yes
Aluminum Sulfate	Sat'd	75	yes	yes	yes
Alums	Sat'd	75	yes	yes	yes
Ammonium Chloride	Sat'd	75	yes	yes	yes
Ammonium Hydroxide	28%	75	yes	yes	yes
Ammonium Hydroxide	10%	75	yes	yes	yes
Ammonium Sulfate	Sat'd	75	yes	yes	yes
Ammonium Thiocyanate	Sat'd	75	yes	yes	yes
Amyl Alcohol	Any	75	?	no	?
Arsenic Acids	Any	75	yes	yes	yes
Barium Sulfide	Sat'd	75	yes	yes	yes
Black Liquor	Sat'd	75	yes	yes	yes
Benzoic Acid	Sat'd	75	yes	yes	yes
Brass Plating Solution	Any	75	yes	yes	yes
Bromine Water	Sat'd	75	yes	no	yes
Butyl Alcohol	Any	75	?	no	?
Cadmium Plating Solution	Any	75	yes	yes	yes
Calcium Bisulfite	Any	75	yes	yes	yes
Calcium Chloride	Sat'd	75	yes	yes	yes
Calcium Hypochlorite	Sat'd	75	yes	yes	yes
Carbonic Acid	Sat'd	75	yes	yes	yes
Casein	Sat'd	75	yes	yes	yes
Castor Oil	Any	75	yes	yes	yes
Caustic Soda	35%	75	yes	yes	yes
Caustic Soda	10%	75	yes	yes	yes
Caustic Potash	35%	75	yes	yes	yes
Caustic Potash	10%	75	yes	yes	yes
Chlorine Water	Sat'd	75	yes	yes	yes
Chromium Plating Solution	Any	75	yes	yes	yes
Citric Acid	Sat'd	75	yes	yes	yes
Copper Chloride (Cupric)	Sat'd	75	yes	yes	yes
Copper CyanidePlating Sol	Any	75	yes	yes	yes
(High Speed)	Any	75	yes	yes	yes
(with Alkali Cyanides)	Sat'd	75	yes	yes	yes
Copper Sulfate	Sat'd	75	yes	yes	yes
Cocoanut Oil	Sat'd	75	yes	yes	yes
Cottonseed Oil	Sat'd	75	yes	yes	yes
Disodium Phosphate	Sat'd	75	yes	yes	yes
Ethyl Alcohol	Any	75	?	no	?
Ethylene Glycol	Any	75	?	no	?
Ferric Chloride	45%	75	yes	yes	yes
Ferrous Sulfate	Sat'd	75	yes	yes	yes
Fluoboric Acid	Any	75	yes	yes	yes
Formaldehyde	37%	75	yes	yes	yes
Formic Acid	85%	75	yes	no	yes
Gallic Acid	Sat'd	75	yes	yes	yes
Glucose	Any	75	yes	yes	yes
Glue	Any	75	yes	yes	yes
Glycerine	Any	75	yes	yes	yes
Gold Plating Solution	Any	75	yes	yes	yes
Hydrochloric Acid	10%	75	yes	yes	yes
Hydrochloric Acid	21.5%	75	yes	yes	yes
Hydrochloric Acid	37.5%	75	yes	no	yes
Hydrofluoric Acid	4%	75	yes	yes	yes
Hydrofluoric Acid	10%	75	yes	yes	yes
Hydrofluoric Acid	48%	75	yes	no	yes
Hydrogen Peroxide	30%	75	yes	yes	yes
Hydrogen Sulfide	Sat'd	75	yes	yes	yes
Hydroquinone	Any	75	yes	yes	yes
Indium Plating Solution	Any	75	yes	yes	yes
Lactic Acid	50%	75	yes	yes	yes
Lactic Acid	Any	75	yes	yes	yes
Lead Plating Solution	Any	75	yes	yes	yes
Malic acid	Any	75	yes	yes	yes
Methyl Alcohol	Any	75	?	no	?
Mineral Oils	Any	75	yes	yes	yes
Nickel Acetate	Sat'd	75	yes	yes	yes
Nickel Plating Solution		75	yes	yes	yes
Nickel Salts	Sat'd	75	yes	yes	yes
Nitric Acid	35%	75	yes	yes	yes
Nitric Acid	40%	75	yes	yes	yes
Nitric Acid	60%	75	yes	no	yes
Nitric Acid/	15%				
Hydrofluoric Acid	4%	75	yes	yes	yes
Nitric Acid/	16%				
Sodium Dichromate	13%	75	yes	yes	yes
Water	71%				
Oleic Acid	Any	75	yes	yes	yes
Oxalic Acid	Sat'd	75	yes	yes	yes
	Any	75	yes	yes	yes
Phenol	Sat'd	75	yes	no	yes
Phosphoric Acid	75%	75	yes	yes	yes
Phosphoric Acid	85%	75	yes	yes	yes
Potassium Acid Sulfate	Sat'd	75	yes	yes	yes
Potassium Antimonate	Sat'd	75	yes	yes	yes
Potassium Bisulfite	Sat'd	75	yes	yes	yes
Potassium Chloride	Sat'd	75	yes	yes	yes
Potassium Cuprocyanide	Sat'd	75	yes	yes	yes
Potassium Cyanide	Sat'd	75	yes	yes	yes
Potassium Diachromate	Sat'd	75	yes	yes	yes
Potassium Hypochlorite	Sat'd	75	yes	?	yes
Potassium Sulfide	Sat'd	75	yes	yes	yes
Potassium Thiosulfate	Sat'd	75	yes	yes	yes
Propyl Alcohol	Sat'd	75	?	no	?
Rhodium Plating Solution	Sat'd	75	yes	yes	yes
Silver Plating Solution	Sat'd	75	yes	yes	yes
Soaps	Any	75	yes	yes	yes
Sodium Acid Sulfate	Sat'd	75	yes	yes	yes
Sodium Antimonate	Sat'd	75	yes	yes	yes
Sodium Bicarbonate	Sat'd	75	yes	yes	yes
Sodium Bisulfite	Sat'd	75	yes	yes	yes
Sodium Chloride	Sat'd	75	yes	yes	yes
Sodium Cyanide	Sat'd	75	yes	yes	yes
Sodium Dichromate	Sat'd	75	yes	yes	yes
Sodium Hydroxide	10%	75	yes	yes	yes
Sodium Hydroxide	35%	75	yes	yes	yes
Sodium Hydroxide	73%	75	no	no	yes
Sodium Hypochlorite	Sat'd	75	yes	?	yes
Sodium Hypochlorite	15%	75	yes	?	yes
Sodium Sulfide	Sat'd	75	yes	yes	yes
Sodium Thiosulfate	Sat'd	75	yes	yes	yes
Sulfuric Acid	15%	75	yes	yes	yes
Sulfuric Acid	50%	75	yes	yes	yes
Sulfuric Acid	70%	75	yes	yes	yes
Sulfuric Acid	98%	75	no	no	yes
Sulfurous Acid	2%	75	yes	yes	yes
Sulfurous Acid	6%	75	yes	no	yes
Tannic Acid	Sat'd	75	yes	yes	yes
Tartaric Acid	Sat'd	75	yes	yes	yes
Tin Chloride Aqueous	Sat'd	75	yes	yes	yes
Tin Plating Solution	Sat'd	75	yes	yes	yes
Triethaneolamine	Sat'd	75	yes	yes	yes
Trisodium Phosphate	Sat'd	75	yes	yes	yes
Water	Sat'd	75	yes	yes	yes
White Liquor		75	yes	yes	yes
Zinc Plating Solution		75	yes	yes	yes
Zinc Sulfate	Sat'd	75	yes	yes	yes

Figure A–14 Chemical resistance chart for Urethane coatings. (Courtesy of Robroy Industries Inc)

PVC Coating Chemical Resistance Chart

Solutions	Conc.	Temp.	Spashing	Liquid	Fumes
Acetic Acid	10%	75	yes	yes	yes
Acid Copper Plating Solution		75	yes	yes	yes
Alkaline Cleaners		75	yes	yes	yes
Aluminum Chloride	Sat'd	75	yes	yes	yes
Aluminum Sulfate	Sat'd	75	yes	yes	yes
Alums	Sat'd	75	yes	yes	yes
Ammonium Chloride	Sat'd	75	yes	yes	yes
Ammonium Hydroxide	28%	75	yes	yes	yes
Ammonium Hydroxide	10%	75	yes	yes	yes
Ammonium Sulfate	Sat'd	75	yes	yes	yes
Ammonium Thiocyanate	Sat'd	75	yes	yes	yes
Amyl Alcohol	Any	75	?	no	?
Arsenic Acids	Any	75	yes	yes	yes
Barium Sulfide	Sat'd	75	yes	yes	yes
Black Liquor	Sat'd	75	yes	yes	yes
Benzoic Acid	Sat'd	75	yes	yes	yes
Brass Plating Solution	Any	75	yes	yes	yes
Bromine Water	Sat'd	75	yes	no	yes
Butyl Alcohol	Any	75	?	no	?
Cadmium Plating Solution	Any	75	yes	yes	yes
Calcium Bisulfite	Any	75	yes	yes	yes
Calcium Chloride	Sat'd	75	yes	yes	yes
Calcium Hypochlorite	Sat'd	75	yes	yes	yes
Carbonic Acid	Sat'd	75	yes	yes	yes
Casein	Sat'd	75	yes	yes	yes
Castor Oil	Any	75	yes	yes	yes
Caustic Soda	35%	75	yes	yes	yes
Caustic Soda	10%	75	yes	yes	yes
Caustic Potash	35%	75	yes	yes	yes
Caustic Potash	10%	75	yes	yes	yes
Chlorine Water	Sat'd	75	yes	yes	yes
Chromium Plating Solution	Any	75	yes	yes	yes
Citric Acid	Sat'd	75	yes	yes	yes
Copper Chloride (Cupric)	Sat'd	75	yes	yes	yes
Copper CyanidePlating Sol (High Speed)	Any	75	yes	yes	yes
(with Alkali Cyanides)	Sat'd	75	yes	yes	yes
Copper Sulfate	Sat'd	75	yes	yes	yes
Cocoanut Oil	Sat'd	75	yes	yes	yes
Cottonseed Oil	Sat'd	75	yes	yes	yes
Disodium Phosphate	Sat'd	75	yes	yes	yes
Ethyl Alcohol	Any	75	?	no	?
Ethylene Glycol	Any	75	?	no	?
Ferric Chloride	45%	75	yes	yes	yes
Ferrous Sulfate	Sat'd	75	yes	yes	yes
Fluoboric Acid	Any	75	yes	yes	yes
Formaldehyde	37%	75	yes	yes	yes
Formic Acid	85%	75	yes	no	yes
Gallic Acid	Sat'd	75	yes	yes	yes
Glucose	Any	75	yes	yes	yes
Glue	Any	75	yes	yes	yes
Glycerine	Any	75	yes	yes	yes
Gold Plating Solution	Any	75	yes	yes	yes
Hydrochloric Acid	10%	75	yes	yes	yes
Hydrochloric Acid	21.5%	75	yes	yes	yes
Hydrochloric Acid	37.5%	75	yes	no	yes
Hydrofluoric Acid	4%	75	yes	yes	yes
Hydrofluoric Acid	10%	75	yes	yes	yes
Hydrofluoric Acid	48%	75	yes	no	yes
Hydrogen Peroxide	30%	75	yes	yes	yes
Hydrogen Sulfide	Sat'd	75	yes	yes	yes
Hydroquinon	Any	75	yes	yes	yes
Indium Plating Solution	Any	75	yes	yes	yes
Lactic Acid	50%	75	yes	yes	yes
Lactic Acid	Any	75	yes	yes	yes
Lead Plating Solution	Any	75	yes	yes	yes
Malic acid	Any	75	yes	yes	yes
Methyl Alcohol	Any	75	?	no	?
Mineral Oils	Any	75	yes	yes	yes
Nickel Acetate	Sat'd	75	yes	yes	yes
Nickel Plating Solution		75	yes	yes	yes
Nickel Salts	Sat'd	75	yes	yes	yes
Nitric Acid	35%	75	yes	yes	yes
Nitric Acid	40%	75	yes	yes	yes
Nitric Acid	60%	75	yes	no	yes
Nitric Acid/ Hydrofluoric Acid	15% 4%	75	yes	yes	yes
Nitric Acid/ Sodium Dichromate Water	16% 13% 71%	75	yes	yes	yes
Oleic Acid	Any	75	yes	yes	yes
Oxalic Acid	Sat'd	75	yes	yes	yes
	Any	75	yes	yes	yes
Phenol	Sat'd	75	yes	no	yes
Phosphoric Acid	75%	75	yes	yes	yes
Phosphoric Acid	85%	75	yes	yes	yes
Potassium Acid Sulfate	Sat'd	75	yes	yes	yes
Potassium Antimonate	Sat'd	75	yes	yes	yes
Potassium Bisulfite	Sat'd	75	yes	yes	yes
Potassium Chloride	Sat'd	75	yes	yes	yes
Potassium Cuprocyanide	Sat'd	75	yes	yes	yes
Potassium Cyanide	Sat'd	75	yes	yes	yes
Potassium Diachromate	Sat'd	75	yes	yes	yes
Potassium Hypochlorite	Sat'd	75	yes	?	yes
Potassium Sulfide	Sat'd	75	yes	yes	yes
Potassium Thiosulfate	Sat'd	75	yes	yes	yes
Propyl Alcohol	Sat'd	75	?	no	?
Rhodium Plating Solution	Sat'd	75	yes	yes	yes
Silver Plating Solution	Sat'd	75	yes	yes	yes
Soaps	Any	75	yes	yes	yes
Sodium Acid Sulfate	Sat'd	75	yes	yes	yes
Sodium Antimonate	Sat'd	75	yes	yes	yes
Sodium Bicarbonate	Sat'd	75	yes	yes	yes
Sodium Bisulfite	Sat'd	75	yes	yes	yes
Sodium Chloride	Sat'd	75	yes	yes	yes
Sodium Cyanide	Sat'd	75	yes	yes	yes
Sodium Dichromate	Sat'd	75	yes	yes	yes
Sodium Hydroxide	10%	75	yes	yes	yes
Sodium Hydroxide	35%	75	yes	yes	yes
Sodium Hydroxide	73%	75	no	no	yes
Sodium Hypochlorite	Sat'd	75	yes	?	yes
Sodium Hypochlorite	15%	75	yes	?	yes
Sodium Sulfide	Sat'd	75	yes	yes	yes
Sodium Thiosulfate	Sat'd	75	yes	yes	yes
Sulfuric Acid	15%	75	yes	yes	yes
Sulfuric Acid	50%	75	yes	yes	yes
Sulfuric Acid	70%	75	yes	yes	yes
Sulfuric Acid	98	75	no	no	yes
Sulfurous Acid	2%	75	yes	yes	yes
Sulfurous Acid	6%	75	yes	no	yes
Tannic Acid	Sat'd	75	yes	yes	yes
Tartaric Acid	Sat'd	75	yes	yes	yes
Tin Chloride Aqueous	Sat'd	75	yes	yes	yes
Tin Plating Solution	Sat'd	75	yes	yes	yes
Triethaneolamine	Sat'd	75	yes	yes	yes
Trisodium Phosphate	Sat'd	75	yes	yes	yes
Water	Sat'd	75	yes	yes	yes
White Liquor		75	yes	yes	yes
Zinc Plating Solution		75	yes	yes	yes
Zinc Sulfate	Sat'd	75	yes	yes	yes

Figure A–15 Chemical resistance chart for PVC coatings. (Courtesy of Robroy Industries Inc)

Index

AC cable, 146
Accessibility, 13
AC voltage, xi, xii, 147
AHJ. *See* Authority having jurisdiction
Aircraft hangars, 85–89
Air spaces, 144–147, 157
Allied Tube and Conduit, 151
Allowable conductor fill requirements, 147
Aluminum, 137
Aluminum metal conduit (ARC), 149–151
Aluminum rigid metal conduit, 150
American Insurance Association, xii
American National Standards Institute (ANSI), 1, 3–4, 161
American Petroleum Institute (API), 3, 33, 161
American Society of Testing and Materials (ASTM), 3
Ampere, Andre Marie, xi
Anesthetizing locations, 113–114
Annular space fillers, 144
ANSI. *See* American National Standards Institute
API. *See* American Petroleum Institute
ARC. *See* Aluminum metal conduit, 149–151
Arcs. *See* Sparks and arcs
Article 505, 117–121
ASTM. *See* American Society of Testing and Materials
ASTM E119. *See* Standard Method of Fire test for Building Construction and Materials
Authority having jurisdiction (AHJ), 4, 6, 13, 24, 33, 161

Basic design factors, 13–19
Battery-charging equipment, 72, 82, 87–88
Bending, 158
Black gas pipe, 151
BOCA. *See* Building Officials and Code Administrators International
Bonding requirements, 64, 67, 73
Branch circuits, 114
Brown's Ferry Nuclear Plant, 143
Building codes, 1, 3, 18, 85, 142
Building Materials Directory (Manila Book), 142
Building Officials and Code Administrators International (BOCA), 17, 142, 161
Bulk storage plants, 99–104
Burial requirements. *See* Underground requirements
Bushings, 172

Cadmium, 138
Canadian Standards Association (CSA), 27
Capacity, 13, 14, 17
Chemical Resistance Chart for PVC Coatings, 181
Chemical Resistance Chart for PVC Conduit and Tubing, 179
Chemical Resistance Chart for Urethane Coatings, 180
Chicago Board of Fire Underwriters, xxii
Chicago World's Fair, xi, xii
Circuit conductors, 114
Circuit disconnects, 96
Class I, 29, 37–57
Class II, 29–30, 59–69
Class III, 30–31, 71–75
Classes, 22, 23–24
Classification of locations, 14, 24–25, 32–34. *See also* Hazardous locations
CMPs. *See* Code-making panels
Coated rigid conduit, 153–155
Coating processes, 105–112
Code of Federal Regulations (OSHA), 7–9
Code-making panels (CMPs), 2
Codes and standards, 1–11
Colombian Exposition of 1892, xi, xii
Color-coding, 151
Combustibility of raceway materials, 144–146
Combustible dusts, 14, 21, 22, 23, 34
    Class II, 59, 60
    electrically conductive, 59, 60, 62, 64, 67
    list of, 167–168
    magnesium and aluminum, 60, 62, 64
    removable systems for, 62
Combustible penetrations, 143
Commercial garages, 77–83, 91, 92
Communication devices, 65, 72
Condensation, 130, 133
Conductor, xi
Conductor insulation, combustibility of, 144–146
Conduit box, 63
Consensus process, 1, 3, 161

Construction sites, 7
Continuity, mechanical and electrical, 136–137
Corrosion protection, 100, 102, 130, 138–142. *See also* Physical protection
CSA. *See* Canadian Standards Association

DC voltage, xi, xii, 147
De Magnete (Gilbert), xi
Designing engineer, 13
Diesel fuel-dispensing areas, 91
Dipping processes, 105–112
Divisions, 22, 23–24, 31, 34
    Class I, 38–47
    Class II, 60
Drainage, 55
Ducts. *See* Plenums
Dust-ignition-proof apparatus, 28, 62, 67, 68, 72
Dusts. *See* Combustible dusts, 14

Earthquakes, 133
Edison, Thomas, xi, xii, 147, 148, 151
Electrical AC and DC formulas, 173–174
Electrical conductivity, 31
Electrical Constructions Material Directory (Green Book), 6, 9, 138, 161
Electrical metallic tubing (EMT), 127–129, 151, 152–153
Electrical Metallic Tubing (EMT) Specifications, 171
Electrical symbols, 175–178
Electrical utilities, 5
Electricity, history of, xi–xii
Electric per se, xi
Electromagnetic fields, 137
Electro-plating, 139, 140, 149, 150
EMT. *See* Electrical metallic tubing
Enamel, 138, 149
Enclosures, 4, 24, 62, 63
Environmental air spaces, 144–147, 157
Equipment
    in aircraft hangars, 86–87
    in bulk storage plants, 102–103
    Class I, 42–49
    Class II, 62–64, 66
    Class III, 72
    in commercial garages, 79–82
    explosion-proof, 48
    in spraying, dipping, and coating processes, 107–111
    in zone concept, 118, 119
Equipment-grounding conductors, 15, 16, 17
Erickson union, 153
ETL Testing Laboratories, 6
European electrical standards, 4
European zone concept, 117–121
Expansion couplings, 135
Expansion and growth, 13, 14, 17, 147
Expansion properties, 130–135

Explosion-proof apparatus, 28, 32, 48, 63
Explosions, 21, 22–23, 25

Factory Mutual (FM), 6, 27
Federal Register, 9
Fibers. *See* Ignitable fibers or flyings
Fine Print Notes (FPN), 161
Fire codes, 1, 3
Fire marshals, 13, 24
Fire Resistance Directory 1991 (Orange Book), 142, 143
Fire-resistive assemblies, 142–144
Fire-resistive construction, 18
Fires, xi, xii, 17–18, 21, 22, 25
Fire/smoke dampers, 146–147
Firestop methods, 143
Fire test procedures, 18, 142, 161
Fire zones, 4
Flammable fuel, 77, 78
Flammable substances, 14, 21, 22, 23, 34
    gases or vapors in Class I, 37, 38
    lists of, 164–168
Flash fires, 71
Flash points, 25, 161
Flexibility, 13, 17
Flexible conduits, 50, 51, 67
Flexible coupling, 38
Flyings. *See* Ignitable fibers or flyings
FM. *See* Factory Mutual
FPN. *See* Fine Print Notes
Franchised facilities, 78
Franklin, Benjamin, xi
"F" rating, 143
Future expansion. *See* Expansion and growth

Galvanized rigid conduit (GRC), 149–151, 161
Galvanized Rigid Conduit Specifications, 170
Galvanizing, 138, 139, 140, 149
Garages. *See* Commercial garages
Gasfitter craftsmen, 148
Gasoline, 25
Gasoline dispensing and service stations, 23, 79, 91–98
Gas tight, 43, 47, 48
General Electric, xi
General Information for Electrical Construction (White Book), 6–7, 138, 140, 141, 162
Gilbert, William, *De Magnete,* xi
Grain elevators and silos, 9, 59, 60
GRC. *See* Galvanized rigid conduit, 149–151
Green Book. See *Electrical Constructions Material Directory*
Grounding, 64, 67, 73
    in aircraft hangars, 87, 88
    in bulk storage plants, 103
    in commercial garages, 82
    in spraying, dipping, and coating processes, 111

Ground joints, 53
Groups, 22, 24, 31, 34
    Class I, 37–38
    Class II, 59–60
    in zone concept, 118

Hazardous atmosphere, 161
Hazardous Location Equipment (Red Book), 7, 162
Hazardous locations, 13, 14, 21–35
    aircraft hangers, 85–89
    bulk storage plants, 99–104
    Class I, 37
    Class II, 59
    Class III, 71–72
    commercial garages, 77–83
    European zone concept, 117–121
    gasoline dispensing and service stations, 23, 79, 91–98
    health care facilities, 113–115
    and *National Electrical Code (NEC),* 28–32
    spray applications, dipping and coating processes, 105–112
    wiring in, 127–160
Hazardous Substances Combustible Dusts, 167–168
Hazardous Substances Flammable Gases and Vapors, 164–166
HCL. *See* Hydrogen chloride gas, 145
Health care facilities, 113–115
Heating, ventilating, and air-conditioning systems (HVAC), 146–147
Heat producing equipment, 25, 62, 72
Hertz, Heinrich, xi
High-rise structures, 18, 142
History
    of electricity, xi–xii
    of raceway systems, 147–149
Hose stream test, 144
Hospitals. *See* Health care facilities
Hot-dipping, 139, 140, 149, 150
HVAC. *See* Heating, ventilating, and air-conditioning systems, 146–147
Hydrogen chloride gas (HCL), 145

ICBO. *See* International Conference of Building Officials
ID. *See* Internal diameter, 149
Identification marks, 8. *See also* Labels; Trademarks
IEC. *See* International Electrotechnical Commission
IEEE. *See* Institute of Electrical and Electronic Engineers
Ignitable fibers or flyings, 21, 22, 23, 31, 71
Ignition, electrical sources of, 25
Ignition temperature, 31
IMC. *See* Intermediate metal conduit
Induction heating, 137–138
Institute of Electrical and Electronic Engineers (IEEE), 3, 4
Insurance requirements, 5, 13, 24, 31, 33
Integral coupling, 153, 154

Intermediate metal conduit (IMC), 127–29, 151–52, 153–55
Intermediate Metal Conduit (IMC) Specifications, 170
Internal diameter (ID), 149, 151, 152, 161
International Conference of Building Officials (ICBO), 17, 142, 161
International Electrotechnical Commission (IEC), 4, 161–162
International zone concept, 117–121
Intumescing, 143

Janitorial maintenance, 34
Jet fuel bulk fuel station, 87
Joints, 53
Junction boxes, 4

Kearney Laboratories, 17
Kerosene, 25
Kwik-Couple, 153

Labels, xii, 6, 7, 25–27, 162. *See also* Identification marks; Trademarks
Life Safety Code, 3
Lighting
    in aircraft hangars, 87
    for Class II locations, 64
    for Class III locations, 73
    in commercial garages, 79, 80, 81, 82
    explosion-proof, 48
    in health care facilities, 114
    portable reel-lights, 49
    PVC coating, 154
    in refineries, 43
    in spraying, dipping, and coating processes, 108, 110
Lightning protection, xi
Limited care facilities, 113–115
Liquidtight flexible metal conduit, 51, 67, 155–56
Liquidtight flexible nonmetallic conduit, 51, 67
Listing marks. *See* Labels
Local codes, 4–5, 31, 33
Locknuts, 172
Loyd, Nancy, xvii
Loyd, Richard E., xvii
Lubrication pits. *See* Pit areas
Lubrication and service rooms, 78, 91, 92

Magnesium and aluminum dusts, 60, 62
Manila Book. See *Building Materials Directory,* 142
Marine areas, 91, 102
MC cable, 50, 55, 146
Membrane penetration, 142
Merrill, William, xii, 147–148
MET. *See* Metropolitan Electrical Testing
Metallic wiring, 4, 149–156

Metallizing, 139, 149
Metric system, 150
Metropolitan Electrical Testing (MET), 6
MI cable, 51, 67, 146
Mine Safety and Health Administration (MSHA), 4
Mobil servicing equipment, 88
MSHA. *See* Mine Safety and Health Administration

National Academy of Sciences, 3
National Association of Fire Engineers, xii, 148
National Board of Fire Underwriters, xii
National Electrical Code (NEC), xii, xvii, 1–3, 148, 162
   and hazardous locations, 28–32
   objective of, 17
National Electrical Manufacturers Association (NEMA), 4, 153, 162
National Fire Protection Association (NFPA), 1–3, 17–18, 33–34, 162
Nationally Recognized Testing Laboratory (NRTL), 162
NEC. See National Electrical Code
NEMA. *See* National Electrical Manufacturers Association
NFPA. *See* National Fire Protection Association
Noncombustible construction materials, 142
Noncombustible penetrations, 143
Nonconductor, xi
Non electric, xi
Nonferrous metal conduit, 150
Nonincendive circuits, 51, 162
NRTL. *See* Nationally Recognized Testing Laboratory
Nylon, 145

Occidental Coating Company, 154
Occupational Safety and Health Act of 1970 (OSHA), 3, 5, 7–9, 162
OD. *See* Outside diameter, 149
Ohm, George, xi
Orange Book. See *Fire Resistive Directory 1991*, 142
OSHA. *See* Occupational Safety and Health Act of 1970
Outside diameter (OD), 149, 151, 152, 162
Overcurrent protection devices, 4

Panelboards, 4, 21, 22
   Class II, 67
   and expansion, 14, 17
Parking garages, 78
Parts, standards for, 1
Perma-Cote Industries, 154
Physical protection, 100, 127–130. *See also* Corrosion protection
Pit areas, 77, 78
   in aircraft hangars, 86
   in bulk storage plants, 101
   in gasoline dispensing and service stations, 92
Plenums, 144–147, 157

Polyvinyl chloride (PVC), 153–155, 156. *See also* Rigid nonmetallic conduit
Pressure piling, 51, 52
Priestley, Joseph, xi
PVC. *See* Polyvinyl chloride; Rigid nonmetallic conduit
PVC Coatings Chemical Resistance Chart, 181
PVC Conduit and Tubing Chemical Resistance Chart, 179

Raceway systems, 4
   allowable conductor fill requirements, 147
   capacity of, 17
   combustibility of, 144–146
   in commerical buildings, 4
   function of, 127
   history of, 147–149
   and physical protection, 127–130
   recommended uses for, 169
   selection of, 149
   underfloor, 17
Reagents, 138–42
Red Book. See *Hazardous Location Equipment*
Residential and custodial facilities, 113–115
Resorts, 91
Rigid metal conduits (RMC), 127–129, 149–151, 161, 162. *See also* Steel tubular conduits
Rigid Nonmetallic Conduit (Schedule 40 & 80) Specifications, 171
Rigid nonmetallic conduits (PVC or RNMC), 50, 80, 127, 156–158. *See also* Polyvinyl chloride
Risk tolerance, 144
RMC. *See* Rigid metal conduits
R & N Associates Inc., xvii
RNMC. *See* Rigid nonmetallic conduits
Robroy Industries, 153, 154

Safety, 1, 7, 13–14
Safety switches, 4
SBCCI. *See* Southern Building Code Congress International Inc.
Schedule 40 PVC, 156, 171
Schedule 80 PVC, 127, 130, 156, 157, 171
Sealing compounds, 51, 54
Sealing fittings, 51–56
Seals, 31, 32
   in aircraft hangars, 87
   and annular space fillers, 144
   in bulk storage plants, 103
   Class I, 51–56
   Class II, 67–68
   in commercial garages, 80, 81
   and condensation, 133
   in gasoline dispensing and service stations, 95–96
   for physical protection, 130
   in spraying, dipping, and coating processes, 111

Secure requirements, 135–136
Service stations. *See* Gasoline dispensing and service stations
Sewage lift pumping station, 6
Sherardizing, 139, 149
Silicone bronze conduit, 150
Solvent cement, 157–158
Southern Building Code Congress International Inc. (SBCCI), 18, 142, 162
Spacings for conduit and tubing, 172
Sparks and arcs, xi, 21, 22, 25, 66, 72, 148
Spraying processes, 105–112
Stanchions, rostrums, and docks, 87
Standard Method of Fire test for Building Construction and Materials (ASTM E119), 151
Static ground clamp, 50
Steel tubular conduits, 17, 148. *See also* Rigid metal conduits
Support requirements, 135–136
Surge arresters, 64, 67
Switchgear devices, 4, 28

Temperatures, 127, 130, 133, 151, 156, 157
Tesla, Nikola, xi, xii, 147, 148
Testing laboratories, 5–7, 27
Threaded joints, 53
Threaded rigid metal conduit, 50, 103
Thread lubricants, 50
Through-penetration fire-stop systems, 143
Touch-up compounds, 154, 155
Trademarks, 6, 25–27. *See also* Identification marks; Labels
"T" rating, 143–44

UL. *See* Underwriters Laboratories
Underground requirements, 130, 131, 132, 155–157
Underwriters Laboratories (UL), xii, 1, 3, 4, 5–7, 27, 148
U.S. Bureau of Mines, 3

United States electrical standards, 4
Urethane Coatings Chemical Resistance Chart, 180
Utilities, electrical, 5

Ventilation, 34, 72
　in aircraft hangars, 86
　in bulk storage plants, 100, 101
　in Class I Division 2, 42
　in commercial garages, 78, 79
　in spraying, dipping, and coating processes, 107, 110
Volta, Alessandro, xi
Voltage regulation, 13

Watt, James, xi
Westinghouse, George, xi, xii, 148
Wet locations, 138, 140, 153, 157
White Book. See *General Information for Electrical Construction*
White rust, 141
Wiring, 127–160
　in aircraft hangars, 86–87
　in bulk storage plants, 102–103
　Class I, 49–51
　Class II, 64, 67
　Class III, 73–74
　in commercial garages, 79–82
　installation in hazardous locations, 31
　metallic, 149–156
　recommended uses for, 169
　in spraying, dipping, and coating processes, 107–111
　in zone concept, 119

X-ray equipment, 114

Zinc coating, 138, 139, 140, 149
Zone concept, 117–121